CLOCKS
REPAIR AND MAINTENANCE

By the same author

Repairing Antique Clocks

CLOCKS
REPAIR AND MAINTENANCE

Eric Smith

Arco Publishing Company Inc.
New York

Published by Arco Publishing Company, Inc.
219 Park Avenue South, New York, N.Y. 10003

Copyright © 1977 by Eric Smith

Printed in Great Britain

Library of Congress Cataloging in Publication Data

Smith, Eric, 1940–
Clocks and clock-repairing

Includes index
1. Clocks and watches–Repairing and adjusting
I. Title
TS547.S58   681'.113'028   76-53567
ISBN 0-668-04187-0

CONTENTS

INTRODUCTION

'Cases of enterohorologomania, as we medicos call it, are rare but not uncommon ... Treatment for the condition is prolonged and not always successful. I always recommend plenty of rest; and, above all, keep the sufferer away from clocks.' So Peter Simple of *The Daily Telegraph* commenting on a curious series of burglaries in eastern France in which the internal organs of clocks were removed but their cases left behind.

This book is for sufferers from some such incurable disease though not, it is hoped, of criminal intent. There are many people who cannot resist clocks. Their primary interest may be mechanical or aesthetic. They may veer towards collecting the old or unusual or to repairing anything and everything that comes their way, having no strong interest in hoarding the result. Often these interests coexist and the collector will both seek to improve the condition of pieces in his own collection and look out for mechanical and other oddities, and also repair the purchases or ordinary chattels, down to the humble 30-hour alarm, of friends. He is fascinated by the life within a going clock and he cannot help but relish restoring life to that which is dormant.

I have tried to write a general handbook to the wide field of broken-down domestic time-keepers which invite attention, to outline the common principles of operation and of the adjustments and repairs which are frequently needed. It is hoped that the reader, by considering for example a difficulty in one striking mechanism in the context of other striking mechanisms, may both find some help for his particular problem and also be empowered to discover things for himself.

This book is therefore divided into sections on the main areas of clock-work rather than looking in detail at several different types of clock, none of which may happen to be the one which has immediately

6

to be worked on. Inevitably, there is some overlapping of sections, so a detailed index has been made to assist reference. The variety of electric clocks is so great, and the relation of the various modern 'electronic' clocks to traditional mechanical horology is so open to discussion, that I have regretfully had to exclude comment on electrical time-keepers. The practical clock-man is, however, rather poorly served with basic information on the working of modern electric clocks and it is to be hoped that before long a book on this subject may appear.

I have assumed that the reader will have some manual tools and materials, but not a lathe or specialist rotary equipment. Many excellent textbooks on clocks assume the possession of (and skill to use) more elaborate apparatus than is, I believe, widely available to the interested amateur. Whilst those without them may well come to invest later in such tools, there is a very great field of interest and repairs open to them before they do so. The basic needs in this direction can be checked from several manuals, which also set out the procedures for dismantling, cleaning and assembling, which I have not discussed here. Some of these matters are also considered in my own *Repairing Antique Clocks* (David & Charles, 1973).

1 THE MOTIVE POWER

For practical purposes, the motive power of domestic clocks may be classified as the falling weight, the coiled spring and electricity. There have of course been variations and unusual combinations. Water has powered usable clocks. Light charging a photo-electric cell has produced the electricity to drive a motor to wind up a small mainspring. In a very common modern combination, a battery and an electro-magnet repeatedly wind a spring. In a range of American clocks of the early nineteenth century, now highly valued as antiques, a flat 'wagon spring' was employed. But now, as for centuries, it is generally true that the chief motive power is the coiled spring or the falling weight whilst in the last seventy years electricity has increasingly come into its own. Electricity is used sometimes to drive dials which are little more than electric motors, sometimes to set up the two main conventional sources of power and sometimes, by the electro-magnet or induction coil, to vibrate a balance wheel or pendulum which drives a gear train.

The uninitiated person may think it naive to ask what is the use to which the power in a clock is put, but after he has wound a manually wound turret clock it may seem a more reasonable question, and to the clock-lover it is fundamental. A great deal of energy is expended in winding a clock relative to the small movements of its balance wheel or the apparent lightness of its hands. A large amount of that energy is expended in friction between moving parts, friction which is kept to the minimum because it is not constant either from day to day in terms of the atmosphere and in proportion to the amount of energy left which has still to carry on doing the same job until the clock is unwound, or from year to year in terms of the irregular wearing away of minute portions of the metals involved. Most critically, little energy is expended in driving the hands round for these, though they may be

9

carefully counterbalanced to minimise the effects of gravity as their tips move upwards and downwards, are in most clocks mere addenda, attached relatively near to the source of power. The energy is expended rather in keeping going the means of controlling and regulating its dissipation.

This means of control is the pendulum or balance wheel, exercising itself through the escapement. It is possible for the two jobs of turning round the hands and keeping the oscillator in motion to be performed from different sources of power, but in most domestic clocks they are performed by one and the same; the spring not only turns the wheels, whose progress is regularly interrupted by the oscillator, but it also gives the oscillator a push, an impulse, usually at every interruption. The frequency of these tiny impulses consumes a large amount of energy, as will be realised if the pendulum of a longcase clock and the driving weight of the same clock are held in the hand and compared.

Thus there is, as it were, a constant battle between the two ends of the clock movement; the more relatively heavy is the oscillator, the better will be the control which it exercises on the unwinding energy, but the greater will be the force required to keep it going and the greater correspondingly will be the friction and wear. The ideal is the greatest mass in pendulum bob or balance wheel which can be sustained in motion by the weakest spring or lightest driving weight. Less critically, there is the same conflict in the striking train. At one end of the movement is the mass of the hammer needed to produce a certain sonority, and at the other is the weight or the spring required (usually) to raise it against gravity and a spring. The smaller the power required to overcome inertia, gravity and the friction of the train, the more efficient the movement.

Weight Drive

Whilst there have been novelties in which the movement itself forms the driving weight, rolling down an inclined plane or creeping down a vertical column until it is rewound by returning it to its starting point, only two falling weight systems have ever been in general use, and they were both extensively employed in longcase clocks. The first and simplest consists of a weight suspended from gut or, more recently, steel cable, which is fixed to an arbor at one end and turns the arbor as the weight falls and the cable unwinds. In practice, of course, little traction is to be had from a line on an arbor, and the gut or wire is always wound round a cylindrical barrel, which may be grooved to keep the turns in place. The duration of the run is doubled by fixing the other end of the line, usually to the seatboard of the movement, and having the weight suspended from the looped line on a free pulley. Approximately double the single line weight is then needed. This duration depends on the fall available for the weight and also on the gearing between the barrel and the centre wheel on which the minute hand turns; since that wheel must revolve once in an hour, the number of turns it makes is directly related to the number of turns which the barrel can make in a certain period. In practice it was usually arranged that in an 8-day clock the barrel revolved some sixteen times

The endless chain

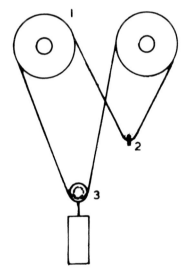
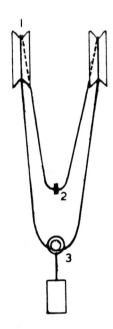

Fig 1

1. Movement's pulleys for rope or chain
2. Counterweight (lead ring through which rope or chain passes)
3. Pulley from which driving ring is hung

11

and the weight fell some 5ft, the striking being designed to correspond.

The alternative arrangement is that of the 'endless chain' or rope. Here the line does not uncoil from a barrel but runs round pulleys which revolve as it passes over them (Fig 1). There are spikes on the pulleys to coincide with the links of the chain or sharper spikes to catch on the rope line, although in some old wooden clocks reliance was had on the friction of the rope running over a narrow wooden pulley without spikes. A single rope or chain is normally used for both striking and going trains of the clock, although it is possible to run a silent timepiece with an endless chain by using a second, idler, pulley mounted in the case. The great advantages of the endless chain are simplicity, reduction in overall weight (since only one weight and a small counterweight are required) and the fact that even during the winding tension continues to be placed on the train (whereas with a simple single line and barrel the train is forced into reverse as the clock is wound, causing inaccuracy and the chance of damage to the escapement). The disadvantages lie in the pulley action and in the strain which a heavy weight imposes on the system if the clock is to run for a long period. At its best the endless chain is liable to slip from time to time on its pulleys, although in the longer term it is less subject to wear than is the rope. In consequence, the system was used on clocks with shorter durations, 30 hours rather than 8 days, since these movements with shorter gear trains have a lesser power requirement. There are 8-day chain-driven longcase movements but they are very rare.

Spring Drive

In one respect the falling weight has no equal as a power-supply for mechanical clocks, and that is in the constancy with which it releases energy from start to finish. Perhaps few families have not had their notorious spring-driven clock which runs fast at the start of the week, slow at the end, and is skilfully adjusted so that the time is nearest right in the middle. There is, in the well-designed movement correctly set up, no necessity for these extremes. But, though measures have been found to counteract it, this inconstancy is an inherent drawback to the coiled spring as a source of power in clocks. The other drawback is the practical one that a spring will eventually break, as will a weight line. When a line breaks, the weight crashes down, ideally onto preplaced padding at the base of the clock, and the energy is dispersed almost entirely in the floor-boards, but when a mainspring breaks, the energy is released into the movement – arbors are displaced and the teeth of gearwheels stripped. On the other hand, the spring has for many circumstances, especially nowadays, the overriding advantage of being portable. Consequently there have been for many centuries portable clocks driven by springs, and static clocks powered by weights. More recently the static clock has tended to be that plugged into the electric mains, but the spring-driven clock, while it may be wound by a battery, has only recently begun partly to be displaced by an economic portable electronic timepiece.

Two devices have been generally used to counteract the diminution of power from an unwinding spring. The first is the fusee, which is in effect a continuously variable gear operating to the disadvantage of the spring when the latter is fully wound, but assisting with leverage as the spring runs down. It takes the form of a grooved barrel, smaller at one end than at the other, connected to the barrel housing the spring by a gut or steel line or by a steel chain. One end of the spring is on a fixed arbor and the other attached to the barrel. As the fusee is wound, the line is unwound from the spring's barrel, which turns and winds the spring, and collected onto the fusee whose base is the first wheel in the gear train. Thus, when the clock is fully wound, all the line is on the fusee save for a length linking the spring with the topmost, narrow end of the fusee (Fig 2). As the spring then turns the fusee, and the gear train, the line is wound back off the fusee in increasingly large circles and onto the barrel. Advantage is taken of the line's rising up the fusee as the clock is wound, so that, as winding is nearly complete, the line raises a sprung finger into the path of a stop piece or 'poke' fastened to the top of the fusee; when finger and poke collide the clock can be wound no further and so there is no chance of overwinding the powerful spring. It is also arranged that even when the clock is unwound, there remains some tension on the line to ensure that it does not wander from the correct path on the fusee. This is done by mounting the spring barrel's arbor in a stout ratchet with a movable pawl. This pawl is only moved when the fusee is initially 'set up'. At this stage all the line is held wound onto the mainspring barrel, the line is hooked or knotted onto the fusee and the spring is wound up a turn by its arbor, the pawl then being allowed to lodge in the ratchet and hold the spring partly wound.

A fusee fully wound   **Fig 2**

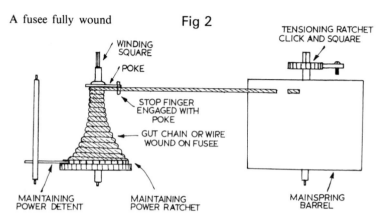

The other device for stabilising a spring's release of power is a means of holding the spring partly wound so that its weak terminal coils do not drive the clock. This is the piece known as a 'going barrel', being a cylinder containing the spring and having gear teeth around its base which makes it the first wheel in the gear train (as is the fusee in a fusee clock). The design of the barrel has to restrict the possible expansion of the spring so that the weak final coils are not used, and equally, of course, it must allow enough room for sufficient turns of the more effective part of the spring to keep the clock going for the desired duration. This is not strictly speaking a matter of hit and miss; there are several variables involved and we shall consider them in due course.

Maintaining power, stopwork, clickwork

It has been said that one of the advantages of the 'endless chain' is the fact that power is not taken off the movement as it is wound. This interruption occurs with the other weight-driven movements and also with fusee springs, but not with going barrels. The problem is not only

14

Maintaining power for weight drive and fusee　　**Fig 3**

Weight drive

Maintaining spring passes through ratchet, on which it is mounted, and engages spoke of greatwheel at x

1. Greatwheel
2. Maintaining ratchet
3. Maintaining spring
4. Click and spring
5. Clickwheel
6. Maintaining detent
7. Spring clip

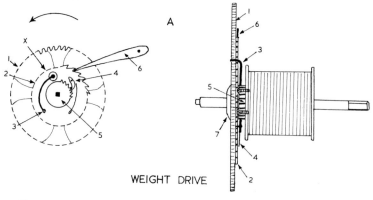

A

WEIGHT DRIVE

Fusee

The maintaining ratchet (2) fits above the greatwheel (3), and both wheels ride freely on the arbor of the fusee (1), being connected only by clickwork and maintaining spring (E). This spring is fixed at one end (C) to the maintaining ratchet and a post (D) at the other end locates in a hole in the greatwheel. 'F' is the fusee click working on the ratchet (G) screwed to the end of the fusee. During winding (here shown clockwise) the maintaining ratchet is held by the detent (A) and the train revolves under the tension of the maintaining spring.

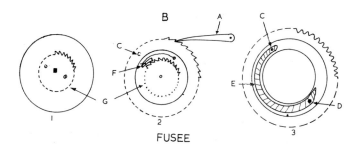

FUSEE

a small loss of time but also the likelihood that the scapewheel will run backwards and be damaged by the pallets during winding. The difficulty is overcome by filling the interval with power derived from a subsidiary spring held in tension by the main source of power. The subsidiary device is known as 'maintaining power' and there are several forms. One of the best known on antique clocks is 'bolt and shutter' where the key cannot be inserted in the movement until a lever has been operated by pulling a cord. This exposes the winding squares and sets up the maintaining power spring. This is a sought-after rarity in old clocks and is seldom found in original form.

More common is the 'detent' arrangement on fusee clocks and high-grade pendulum clocks (Fig 3). Here, at the base of the fusee or back of the barrel, just above the going teeth, is a ratchet wheel with fine teeth, into which drops a long finger or detent. This ratchet is solid with a circular spring which acts on a hole in the greatwheel (that is, the foot of the fusee) or on a spoke of the weight barrel. When the clock is under power this spring is tensioned and the detent lodged behind a ratchet tooth, and, when power is removed by the act of winding, the tension of the spring expends itself against the detent, attempting to push round the ratchet and therefore the barrel or fusee to which the ratchet is attached. The subsidiary spring will operate the clock for two or three minutes, more than sufficient for winding of the main weight or spring. It must, however, be relatively weak or the power of the weight or mainspring would not be able to keep it set up.

The danger, not to mention the inconvenience, of breaking a mainspring through overwinding has been guarded against in many ways, classified under the broad head of 'stopwork'. Stopwork has also the advantage of cutting out the less effective early turns of a spring. Perhaps the commonest of all forms (apart from that used on fusee movements) is 'starwheel stopwork'. It consists of a circular wheel of which a section is replaced by a projecting finger (Fig 4). This part, the stop finger, is fixed rigidly to the barrel arbor. On the barrel is screwed a 5-toothed wheel resembling a Maltese cross. Its teeth all have concave tips save for one, which is convex, and between them are square notches. In operation, the starwheel is turned forwards by the stop finger one notch for each revolution of the

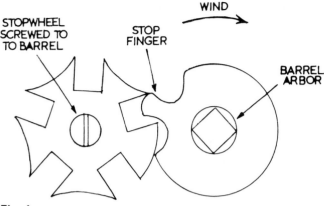

Fig 4

Starwheel stopwork (shown with spring fully wound)

winding square. The rounded tooth of the wheel, however, butts onto the stop finger piece after five revolutions so that the clock can be wound no further; conversely, the wheel revolves against the stationary finger as the clock unwinds, and it can make no more than

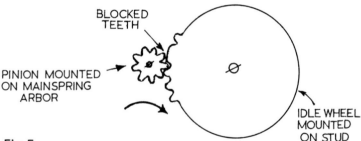

Fig 5

Simple stopwork (shown with spring fully wound)

five revolutions. Other types of stopwork also utilise wheels of irregular shape and the principle that a static finger will pass concave but not convex surfaces (Fig 5).

Whether the power is by weight or by spring, the barrel or fusee is free, restrained only by the operation of a pawl and a ratchet, the 'clickwork', the ratchet being of course passed when the clock is being wound, but catching the pawl or click when the power is acting on the movement. The click, in fact, takes the full force of the power-supply and has an importance out of all proportion to the treatment it is inclined to receive. Clicks and clicksprings vary considerably (Fig 6). In an 8-day longcase clock the clickwheel is solid with the barrel, the click and its spring being fixed to the greatwheel, the wheel on the barrel arbor. If maintaining power is fitted, the click and spring will be mounted on the ratchet wheel, itself connected by a lug to the greatwheel. The click is usually loose on a brass pivot, as is also the arrangement inside a fusee. The click of a 30-hour longcase clock is of a special form, being a steel ring with a raised nub fastened to the winding pulley. There is no ratchet, as this special click acts noisily upon the spokes of the greatwheel. In more modern versions a steel click is sprung and pivoted onto the face of the pulley, again acting upon the spokes of the wheel. With going barrels the usual arrangement is a clickwheel mounted loosely on the winding square and a click and clickspring, which may be of strip or wire, fixed to either the front or the backplate. Where, in the cheaper movements, there is no barrel and the spring is attached to a post of the frame, it is usual for the clickwheel to be mounted solid with the mainspring arbor, and for click and wire clickspring to be on the greatwheel between the plates.

On many modern clocks the clickwork is mounted on a subplate containing one of the holes for the mainspring arbor – in some chiming clocks all three sets of clickwork are mounted on one subplate. This makes for convenience in assembly and simplifies the repair or replacement of a spring, since it is normally obvious that a spring is broken and the rest of the clock may not (though ideally should) receive attention. It also facilitates adjustment, particularly of the striking and chiming trains, without the complication of the springs which otherwise would have to be laboriously let down. Another method of making the barrel removable from assembled plates is to have it mounted on a hollow arbor as long as the distance between plates, the steel arbor and winding square passing through and being secured by a pin to the hollow barrel arbor. Again, it is

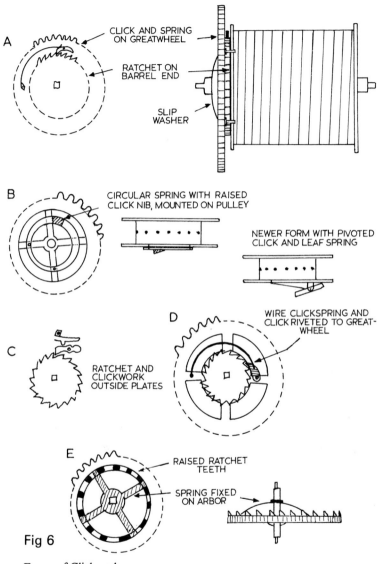

Fig 6

Forms of Clickwork
A   Eight day, weight-driven
B   Thirty-hour, weight-driven
C   Going barrel
D   Open spring
E   Going barrel, four-arm spring with crown ratchet

common for the arbors to run in slots (rather than holes) in the plates, being held in place by screwed straps with holes, one of the straps also serving to keep the clickwheel in place.

Adjustments and Repairs
Weight Drive
Little can go wrong with weight drive, and if there is a breakage the rest of the movement is usually undamaged. The single line breaks most often where it is attached to the barrel. Here a knot is lodged inside the barrel and the line emerges through a hole. It is a simple enough matter to renew the fastening. The other end of the line passes up through a hole or slot in the seatboard and is fastened by a large bow to stop it from slipping back through the hole, or preferably knotted round a short steel pin. More common than an actual breakage is fraying. A frayed line must be replaced, both because it is weak and because loose strands can jam the works. Lines, whether of cable or of gut, become set in their ways. A clock in which the lines are wound onto the barrel crooked cannot run smoothly, and where the lines have been crooked for any time, they will always rewind crooked. This leads to kinks which will produce future breakages. Obviously, judgement must be used according to the circumstances, but it is wise always to replace kinked lines. A new gut line should be oiled before use by securing one end and pulling the line slowly through the fingers with a drop of oil on them. Gut (or nylon) and wire lines are not difficult to obtain, though in an emergency three 'cello A gut strings joined by reef knots will do for a short period. Knots at either end of gut lines may be made to hold by pressing the tip against a hot flat object, which will splay the end. There is no easy way of knotting steel lines, but it is best to form the knot, and then to draw it tight with the metal heated to red heat. If a securing pin is not used, bows and looped knots are necessary at the seatboard end.

The weight pulley should be checked to ensure that it runs freely and its pivots should be given a little machine oil. It is essential that the click action be secure. The ratchet on weight-driven clocks tends not to wear seriously, since the clicksprings are relatively light, but the click itself may need to be filed so that it presents a flat surface square to the wheel teeth, and if it is sloppy on its mounting pin a new pin should be hammered in. The long semi-circular clickspring may be bent slightly inwards to tighten the action and it is not a difficult matter to replace this spring from filed brass, preferably hammered hard, if, as sometimes happens, it is cracked or badly buckled.

Repair work is rarely required to the maintaining power, but the latter must be adjusted to work satisfactorily. This is a matter of ensuring that when the driving weight is hanging from the barrel it causes the maintaining spring to be at tension, and also of ensuring that the detent finger is neither too rounded, nor its pivots too shaky in their holes, for it to latch positively into the teeth of the maintaining ratchet. If the weight (provided it is adequate to drive the clock) will not set up the maintaining power, then the maintaining spring must be weakened by filing a little metal away, or by bending it slightly, according to the type.

The endless chain or rope mechanism, whilst it has advantages as has been said, is often a source of trouble, particularly in the wearing or breaking of the line. Ropes have a fairly short life and many owners like them to be replaced by ropes rather than chains, for the sake of character. Ordinary clothes line will not do, because it is far too stiff. Though there are some forms of nylon rope which are satisfactory they look disturbingly like twentieth-century nylon rather than seventeenth- or eighteenth-century rope. The correct soft, multi-stranded rope is sometimes available but it is no easy matter to join the two ends together in such a way that there is no lump impeding progress over pulleys and a source of future wear. Quite good results can be had from sewing the ends together with linen thread, preferably after they have been divided into at least three strands each and spliced in a rudimentary manner. Ideally, each of the strands, which may number nearly twenty, should be cut to a different length and so knotted with its opposite number that the series of knots takes up some six inches, the whole being oversewn to take in stragglers. It is also possible to stick the ends of the rope together with a resin glue. Sash-cord or picture cord can be used on clocks (usually continental) requiring finer ropes.

Sometimes it may be necessary to follow historical development and replace endless rope by endless chain. For this purpose different pulleys, with blunter and shorter spikes, will be needed and also a chain with links spaced according to the distance between spikes. Ready matched conversion sets can be obtained and with these converting rope to chain drive is not a difficult operation, although it may involve some filing (or preferably lathe turning) of the arbors so that they will go through the pulley bosses. In the typical 30-hour chain clock, one pulley is driven onto the going train arbor, often being pinned through the wheel to keep it firm, whilst the other with the clickwork is loose on the striking arbor and held in place by a washer and a retaining pin right through the arbor. Exactly the same arrangement is followed with the new chain pulleys, washers being added if the strike pulley is loose, or the boss being filed back if it is too long for the pin to be inserted.

It is noticeable that when in 30-hour movements the going and striking trains are side by side between the plates there is very little trouble, but that there is often slipping, even with a new pulley and perfectly matched chain, on the 'bird-cage' movements where the striking train is behind the going train in an iron frame. The reason for this is plainly that the chain can only pass reliably onto the spikes of the pulley if it approaches them in a consistent alternation of horizontal and vertical links. A crooked link is likely sooner or later to foul the side or bottom of the pulley and miss the spikes, when the whole chain may start to slide. The endless chain in the 'front and back' movement has to run fore and aft as well as in line with the pulleys, and so it is prone to slip. With a correct new chain this is an occasional event, but with an old stretched chain (or, as is not infrequent, with the old chain that the dealer happened to have lying around) it is a virtual certainty. If there is no correspondence between spikes and links when they are compared, and if the links are closed, then a new chain will have to be bought. It is wise then to offer the supplier one of the pulleys to ensure that the best fit available is obtained. If, however, the chain is stretched and the links open, it is possible to close them up one by one with pliers, though it is a long job. It need hardly be said that it is essential when fitting the chain to make sure that it runs in the right direction and that the counterweight (which should be of about 6oz) is included in the circuit.

The weight pulley should be checked for freedom and oiled, but often the other really important repair with these clocks is to the clickwork. This, as has been said, takes the form on older clocks of a ring of steel, bent so that it is raised, and with a nub or ledge which engages with the spokes of the greatwheel. These spokes will take, and will receive, a great deal of wear without ill effect, but it is common for the one-piece click and spring to be cracked or broken where it is riveted to the pulley. These pulleys are usually riveted together and it is not too difficult to separate them so that the click rivets can be renewed. The clicks themselves are not objects of beauty or well finished as a rule and it is possible to make replacements from sheet steel, either of thick metal filed down to the right shape, or of thin metal with the nub soldered on. Whether it is worth the trouble, however, is another matter and, depending on the state of the pulleys and the value of the clock, there is much to be said for fitting a new set of pulleys and chain. Clicks of the more modern pattern, with a sprung bar rising to catch the spokes, cause less bother, though here again it will be necessary to take the pulley apart if the spring, which is usually brass strip, has to be replaced and riveted on.

The weights are not items of precision. About 8lb suits most 30-hour clocks and 10lb 8-day clocks (12lb if for the striking train). A Vienna regulator weight may be some 2lb. One always uses as little weight as possible. Weights and pulleys can be bought, but they are expensive. Handsome weights can be made with 2–3in brass tube filled with scrap lead. Pendulum bobs can be made in the same way. The rod should be protected in a metal tube, or may be rubbed with a graphite pencil and greased to prevent it from sticking during the casting. Brass discs can be sweated to top and bottom to finish the job, the bottom hole being drilled when all else is finished.

Spring Drive

With regard to mainsprings, there are one or two rules of thumb which it is useful to know, for instance when one suspects that an incorrect spring has been fitted or when one wishes to replace a spring with one which is 'in stock'. First, it is principally the metal's thickness which determines its strength, given two springs of the same metal. In fact the output varies according to the square of the thickness – if you replace with double-thick metal, you quadruple the power available. Secondly, the last coil of the spring when it is wound should be in the same place in the barrel as the first coil when it is unwound. In other words, to get the maximum number of turns out of the spring in the barrel, the area of the barrel covered by the spring, whether it is up or down, should be approximately equal to the area which is uncovered. Clearly, you must have at least the number of turns available to turn the centre wheel through 30 hours or 8 days, or the specified duration.

Clocks sometimes come to a repairer with the complaint that they stop on the sixth day, or thereabouts, and the inference of the customer is that a stronger spring is needed. This may be so. More often, someone has already come to that conclusion and fitted a spring which is too powerful, too thick and too short, for it is a practice of bodgers to fit a stronger spring when what is really required is some time spent on the escapement and on bushing several of the pivot holes. The spring which is too short and powerful may drive the clock after a fashion for a while but the period of running will be restricted and the clock will start fast and finish slow. Meanwhile widespread damage will have been done throughout the movement and it will require a major overhaul to persuade it to run with a more suitable spring. For uniform power, what is required is a long spring of which the larger coils scarcely come into action. On the other hand, too long a spring will, according to thickness, not be able to be wound a sufficient number of turns to keep the movement going, even if it has enough strength. Therefore a compromise has to be found. One has to remember that there are constraints on the length and thickness of the spring as important though less obvious than the limitation of its height by the height of the barrel.

Springs become tired and set in over-narrow coils and on the whole little can be done besides replacement. It is, however, possible to draw a spring out horizontally, except for the last turn or two, with some benefit. Secondly, a dry spring very soon becomes a rusty spring. A spring oiled with watch or thin clock-oil will soon become dry – especially if the edges are oiled in the hope that some will trickle down.

Springs should be treated with a thin grease, not sticky, but better able to stay in place on a flat surface than fluid clock-oil, and the lubricant should be wiped between the coils.

Broken mainsprings have generally to be replaced. For the amateur, there are, however, sometimes pressing reasons (keeping the bill down, difficulty in obtaining the right spring, the fact that his own clock can always receive attention again) for trying to repair them even at the risk of a later explosion. In general, a spring broken at the centre cannot be repaired, a spring broken in the middle may be but not for long, and a spring broken at the outer end has the best chance.

To clean a spring it is permissible to pull it out gently in a horizontal direction (never vertically in a spiral as this distorts it) and to wipe it with an oily rag between the fingers, but not to extend the two innermost coils, since this could only result in permanent bending, weakness and a crack. For the same reason, even if it were practicable to drill a new hole on the inner end of a spring it would not be desirable. The outer end may, however, be filed off at the corners to a rounded tip, heated to red heat to soften it, then lightly punched and drilled. The final shape of the new hole will depend on the shape of the barrel hook, but it should not have sharp corners and there should be plenty of room for the hole to move on the hook, as it will when the spring is wound. Alternatively, the softened tip may be punched to raise a nipple in the metal. This is then filed off to produce a small hole, which is enlarged by filing. The new hole must be central in the spring and a position from the end rather less than the spring's height is satisfactory; too long a tip will make the spring hard to insert and create a strain at the point of the hole.

The outer ends of springs are hooked in two other common ways and again repair is fairly reliable (Fig 7). The first way is the riveted tag, where a short length of spring is riveted onto the mainspring and pressed behind an inner ring in the barrel edge. Here the old hole has to be broken off and a new tag of spring riveted on. The other connection is by various means of bending. A simple hook may be bent back on the end of the spring and lodge in a hole or slot of the barrel, or a bend forwards may be made, resting round a post formed of two slots or holes in the barrel. Either way, the end of the spring

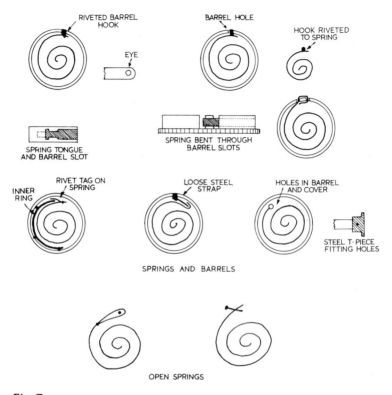

Fig 7

Anchorage of mainsprings (*above*) Springs and barrels (*below*) Open springs

can be softened and bent gingerly with round-nosed pliers to the required shape. The same applies to the end of open springs without barrels. A break in the middle of a spring can sometimes be temporarily (only) repaired by forming a hole on one side, as when making the hole at the outer end, and a tongue on the other. Naturally a compromise must be found in the size of the hole which will not weaken the one part and the need to avoid too thin a neck on the other. The same form of tongue is also sometimes used to hook onto the barrel. Riveting may also be possible, but is not good practice since it introduces a stiff bulge in the coils.

The outer end of the mainspring is caught by the barrel hook, if there is a barrel, or by a pillar or special projection of the frame if there is none, and the inner end is caught by the barrel-arbor hook. The strain on these two hooks is enormous and it is vital that they be firm in their mounting and shaped so as to catch the spring beyond all doubt. In older clocks, the barrel hook was driven, or sometimes screwed, into the barrel. If it is broken or has come loose it is most easily replaced by a hook shaped on the end of a steel screw. The barrel is tapped lightly – allowing for any offset angle there may have been on the previous hook – and the new hook forced home. The end can then be cut off and riveted over from the outside before being filed smooth. If possible, the new hook should be screwed in from inside the barrel, but on small barrels this is not always practicable. Sometimes the hook is in fact riveted to the spring and is a slide-fit into an angled hole in the barrel. It is difficult to replace this type of hook reliably and it is best to fit a new hook and eye of the usual type. The most difficult type of barrel hook to deal with is the most modern and, probably, the cheapest. It consists of a hook pressed up out of the barrel wall, leaving therefore a hole where it has been raised. The attempt to raise another similar hook often results in fracture of the metal and in any case produces an enlarged hole in the barrel. It is safer to rivet a hook into the barrel, preferably in the same place, altering the shape of the spring's eye to fit.

No less important is the barrel-arbor hook. These hooks wear little and they do not break, but they are not always very satisfactory in their original state. It is important here (as with the barrel hook) that the hook is raised above the circle little more than the thickness of the spring. If it is raised much more, it will foul the first coil when the spring is wound and at this point of the spring such a strain easily causes fracture. At the same time, it is vital that the inner end of the spring lodges really securely on the hook. There can be few repairers who do not from time to time have the maddening experience of assembling a complicated clock and winding it up, only to find that the spring is slipping, usually at the inner end. When all has been done to spring and barrel and they are reassembled, it is therefore prudent to test-wind the spring before it is put back into the clock, because an ill fit of the end on the barrel-arbor hook sometimes only shows under pressure. The end must be gradually bent, without sharp angles, until the fit is good.

A barrel is not essential, but if it is used it must function properly. A barrel which wobbles, or of which the top comes off, is a danger to the movement, a waste of power, and liable to impede a neighbouring wheel in the train. In secondhand clocks the most common cause of a loose or floating top on a barrel is too high a spring. It is obvious that the spring used must allow room for the thickness of the barrel top metal inside the barrel, or the spring will rub in a well-shut barrel. If that condition is met and the top is still loose, the turned groove of the barrel may be damaged, in which case it can be straightened with a file (or, better, by turning) or the barrel top has to be stretched by peening with a light hammer. At this stage the top should be tried in position for fit on the mainspring arbor. It requires little shake, but it must, of course, be quite free, and stretching may have tightened the hole. Top and bottom holes are subject to wear and the bottom hole should be tightened if necessary either with a hole-closing (hollow cone) or stout flat punch, being broached to a good fit. If the holes are loose, the barrel will wobble on the arbor and in due course cease to mesh with the first wheel; the spring will then explode and almost certainly damage wheel-teeth. The top and bottom holes are usually in a dome, a reinforcement of raised metal. The size of this dome should be compared with the size of the barrel arbor and hook. If, as occasionally happens, the dome is too large, it will baulk the innermost coils of the spring when wound and lead to jumping of the barrel-arbor hook.

The same applies to the clicks of spring-driven movements as to those of weight-driven ones. It is vital that they be firm and positive in their action and it is folly not to repair them when they are worn. Ratchets and clicks can be bought and often it is hardly worthwhile making them, but in any case it is not too difficult a matter to file them up when required. It is quite common to receive family clocks that have been 'going for years' and to find as first evidence of the fact the deplorable state of ratchet and click. Frequently the click has a hole worn half through it (half, because it was thicker than the ratchet in the first place) and the teeth of the ratchet are burred over so that its ability to catch the click is a miracle. Stiffening the clickspring is no answer here; the only course is replacement of the worn parts. Clicksprings, whether of wire or of strip metal, can be made and bought, but it is not usually they so much as their mounting which cause trouble. It is remarkable how often a clickspring is screwed home in the hope that it will do its arduous and responsible job by virtue of the mere tension of a screw. In practice, of course, it is pushed aside by the first child to 'wind the clock backwards' and even by less specific causes, and one is offered a clock where 'I think the spring must have broken'. Fortunately, as a rule the mainspring has not broken, though some wheel-teeth may have done, but merely jumped its hook. The trouble, however, is, likely as not, caused either by a click whose profile does not match the profile of the ratchet teeth with which it has to engage, or by a clickspring which never had steady-pins or has lost them. From this point of view, particularly in French carriage clocks, where a clickspring has no steady-pin, it is well worth fitting one right away. It can do no harm and it might save a great deal of trouble. The other obvious essential with clicksprings is that they neither rise off the click into the air nor dig down into the metal of the plate beneath them; the spring must deliver its full impact to the back of the click. Similarly, ratchets must be free to turn, but not to rise into less than full engagement with their clicks.

The winding of some modern chiming clocks is through a simple system of gears. This reduces the effort needed to wind powerful springs direct. The clickwork is usually on a wheel mounted on the front plate and through which the mainspring square passes. The winding square itself is fitted to a wheel meshing with this (usually through an intermediary wheel to arrange the same direction of winding).These wheels are subject to wear, but repair is not usually difficult. The winding wheels themselves can be removed without the spring's unwinding, and for a while damage may be made good by inverting them and presenting fresh tooth surfaces. The teeth are few and large; it is possible to file up replacement wheels by hand although, of course, properly cut wheels are preferable.

When it comes to stopwork on the going barrel the repairer – to judge from the number of incomplete clocks one sees – frequently opts out. And perhaps that is understandable. The clock will, after all, function without stopwork. If it is properly designed it will not utilise the last turns of the spring in its normal running period, and the spring itself will give a good indication of when it has been wound far enough. Nonetheless, stopwork is a refinement with good justifications even if it is a refinement. Its difficulty for the repairer lies principally in the irregularity of its parts. The mass-production of stopfingers and 'Maltese cross' wheels is perhaps relatively simple, but these are not shapes that can be run off on a machine by the amateur, and they can only be produced by hand with a great deal of time and trouble. Clearly, it cannot be worth adding stopwork to a clock which does not have it, but whether it is worth restoring missing stopwork must depend on the value of the clock. Adjusting the machinery is a matter of going to work with a file after careful observation. What is essential in any form is that the barrel be entirely unhindered for the prescribed number of turns and definitively stopped when it tries to go beyond that number. The commonest culprit in bad starwheel stopwork is the stopfinger. If the projections on either side of the finger itself are too large, they foul the concave teeth on the wheel, and if the whole fingerpiece is too small it fails to turn the wheel. With these faults, while the winding-up may proceed satisfactorily, the clock may nonetheless be brought to a halt as the fault comes round. The other problem is the strain on the mounting of the stop-wheel, whose screw tends to be forced from the barrel. The only repair is to bush and replace the hole or to enlarge the wheel hole and tap the barrel for a bigger mounting screw – and finding such a screw which will tighten to the correct depth, leaving the wheel free to revolve, is not easy. So that the lower-powered turns of the spring are unused, the mainspring should be partially wound and held at about one and a half turns as the starwheel is screwed to the barrel, its convex tooth engaging with the stop-finger. This setting up must obviously be done at a stage of assembly when the barrel is still accessible.

Winding squares often reach one in a very unattractive state, particularly those of going barrels with their coarse clickwork and direct turning of the spring. This is not only an aesthetic blemish but also inefficient. Burrs should be hammered flat and the square should be reshaped with a fine file, the end of the square being polished for appearance. The object is to remove as little metal as possible during the reshaping because, while it is always possible to change to a smaller key, the key size and square should be proprtionate to the power of the spring. The hammering saves unnecessary removal of metal and also results in a hardened surface. When a square is hopelessly rounded or is broken off short it is possible to fit a replacement. The arbor is drilled deeply to receive a long steel plug, on the end of which the square will be shaped. The rod may be driven home into the arbor or tapped and threaded for a screw fit. Either way, it must be tightly pinned through and the ends of the pin filed flush with the arbor. Care is needed in the operation to secure greatest strength and to avoid splitting the arbor. It is, of course, better if possible to file, or preferably turn, a complete new arbor from steel rod.

If the holes are worn in the plates or in a going barrel or if a fusee line breaks, there is a good chance that the full power of the mainspring will be released, causing teeth of fusee or barrel to be broken. Because of the need for accuracy here for the smooth conveying of power and in view of the strain under which these parts work, there is no doubt that the best course is to have a new fusee or barrel cut. There are times, however, when only one or two teeth are broken and it may be worth trying a repair. The procedure is to flatten the broken teeth and to drill into the thick edge of the barrel or fusee to about a tooth's depth. A piece of steel wire is filed into the rough shape of a tooth and driven into the hole. It may be sweated in with solder for additional strength. The tooth is then filed accurately against the profile of the teeth alongside. Where teeth are very large, it may be preferable to use two wire plugs and smaller holes alongside each other, for each tooth.

A barrel which has revolved at speed against the neighbouring pinion in such cases is often badly worn all round and meshes very insecurely with the pinion when the holes are repaired. There is then little choice but to fit a new barrel. If the spring is not too powerful and the wear not too serious, it is sometimes possible lightly to file the teeth back to a nearly correct profile, removing as little metal as may be, and to arrange reasonable engagement of the pinion by moving the pivot holes very slightly. The holes are likely to have to be drawn and bushed anyway. Occasionally also, a bent tooth can be straightened with a screwdriver as lever, but it will be weakened and may well snap off. These are makeshift repairs and the decision to try them must depend on the nature of clock and customer.

Much of what has been said of springs and going barrels applies in general also to the fusee system. The need for positive action in the clickwork is still more essential because the clickwork is much finer and is invisible when the clock is assembled. The fusee may be connected to the barrel by gut, wire or chain, the latter being reserved for clocks of high quality. Frayed gut and wire must of course be replaced. Gut can be twisted to prevent fraying, as it is wound on. The springs of English bracket clocks are of great power and the lines are fastened to the barrels in a special way, but held in the fusee by a single knot. In the case of a chain, there is a rounded hook to go round a pin in the fusee and an anchor-shaped hook for a hole in the barrel (Fig 8). A damaged or broken chain can be shortened by a link and riveted together without difficulty but if a chain does have (regrettably) to be replaced by gut or wire, it.is merely a question of removing the pin in the fusee and taking the hole through to the inside, and of making two more holes alongside that in the barrel. The question of whether to use gut or wire is really a matter of precedent and preference. Wire is liable to score a barrel and is not easy to handle. On the other hand, it is likely to last longer. In the older clock,

Fig 8

FUSEE   BARREL

Securing fusee chains and lines

particularly where the movement is visible through side panels, gut may be preferred. If a chain is present, it must be relaxed and clean. Rust can be removed by rubbing on fine grade emery on all sides and then brushing. The chain should be soaked in oil for several hours and then cleaned off in paraffin or benzine.

In setting up, the fusee appears more of a complication than it strictly is. This is because, if gut or wire is involved, barrel and fusee have to be connected to the line before the movement is assembled and thus there is line trailing in the way of other operations. (A fusee chain is, however, connected after assembly, since only accessible hooks are involved.) Particular care must be taken to ensure that the line does not become looped round one of the movement's pillars, for it cannot be extricated once the movement is put together. Therefore regard has to be had to the direction of rotation of fusee and barrel. Once the movement is assembled, if there is a chain connected to the barrel, grip the barrel arbor, with pawl slung aside, in a hand vice and wind all the line onto the barrel evenly. The chain can then be hooked onto the fusee. With chain or line connected, then continue turning the mainspring arbor for another three-quarters to one full turn, thus setting up the spring and placing tension on the line to hold it in place. Slide the mainspring's pawl or click into position against the ratchet and screw it home firmly. Finally, test the stopfinger's action by winding up the movement and observing that the line operates the finger and brings it into contact with the fusee poke not more than a turn before the end of the fusee is full. Ensure also that the finger cannot foul the poke at any other time. Sometimes the back edge of the finger, which acts as a stop on its downward movement, has to be stretched with a hammer to prevent finger and poke from engaging too early; or the part of the finger which meets the line can be reduced with a file.

There are two further points to make with regard to springs. The first is that anyone contemplating much repair work is strongly advised to obtain a good mainspring-winder. Springs can be extracted and returned to barrels by hand, but it is an extremely strenuous and rather dangerous business, especially with the spring of an English fusee clock. Furthermore, the necessary spiral motion imposed is very bad for the spring and can cause it to seize on the top and bottom of the barrel. It is also difficult to carry out tests on springs in the wound state without a winder.

The second point concerns the general advisability of dealing with springs. It is a great temptation when doing a hurried repair or cleaning job to leave the spring alone unless it is known to be broken. But it is hoped that this chapter will have shown that a great deal can go wrong with springs, especially with the apparently faithful going-barrel. Pulling them out by hand is certainly a risk and one that has to be balanced against possible good in each case, but on the whole it is only sense to examine the effectiveness of the power-supply before you turn your attention to what it is used for.

2 THE GEAR TRAIN

An ordinary clock is kept to time by its oscillator, be it foliot, balance, pendulum or tuning fork, not by its motor, the train and weight or spring or other source of power. Ideally it will keep the same time whether the arc or swing of the pendulum or swing of the balance wheel is large or small, and regardless of variations in the power delivered to it. In practice this is not quite the case for many reasons, but it remains true that the purpose of the gear train is not in itself to slow the motor nearer to measurable time, but to reduce its power so that it can drive the oscillating controller over a longer period.

The necessity for this reduction to match a manageable size of pendulum (for example) is evident when we contemplate in fantasy a clock with a pendulum whose natural vibration takes a minute. Such a pendulum will be some $2\frac{1}{2}$ miles long. Moreover, it will not divide time into small enough segments to correspond to our sense of continuity; we shall have in exaggerated form the curious speculation as to the precise time indicated by the minute hand of an old longcase clock with a recoil escapement, whose hand steps backward and leaps forward at each second, this being perceptible in the minute divisions of the dial and even more so on the seconds dial. A 1-sec pendulum, however, a mere 39.14in long, could not be driven for any sustained period by the near-direct power of the spring or weight which convenience and custom determine shall operate the clock for 30 hours or 8 days. The pendulum or wheel must be driven by a reduction gear, the barrel revolving only a few times in the duration (normally some 16 times in 8 days) as the scapewheel revolves many thousands of times. If we look at it another way, the oscillator does not normally drive the clock, but its controlling influence is such that it may be imagined to do so. (In some modern electric clocks, however, the oscillator itself does actually drive the clock.) From this point of view the frequent vibrations of the pendulum, once converted into revolutions of wheels, must be reduced until they conform to the standards of a revolution in a minute, an hour and a day. Thus the whole art of clock-gearing lies in efficiently relating the relatively fast-moving oscillator, with its naturally short period of vibration, to the source of power which must keep it in motion over a period.

Similar considerations govern the striking, where a certain interval between blows on a bell is desirable if they are to be distinct, and where a very large number of blows (nearly 1,000 even in a simple hour clock) is to be struck in, say, a week. There is no escapement here, but rather a fly or fan with its inherent inertia modified by the vagaries of climate and friction. The fly further slows down the motor, the mass of whose gearing would of course slow it very considerably without a fan at all. But again we are concerned with exact revolutions, for there is a precise sequence of striking to be repeated every twelve hours (according to the striking and chiming set-up) and it must be let off by a low-powered wheel near the fly when the hammer is correctly placed to strike. In the simplest mechanism of all, the alarm, there is no ratio of gears governed by the need for an exact number of revolutions in a set period, but there is still the need for the alarm to sound for a reasonable time, and this means reducing the power of the spring by simple gearing.

The basic arrangement of trains is that there is a separate train for each function. Thus in a spacious bracket clock there may be a train each for the going, the striking, the chiming and the alarm. Electric clocks based on synchronous motors (which drive the gears from the reverse direction since the natural speed of the motor has to be reduced to that of the clock dials) use a partly common train for going, alarm and striking, and in fact in the field of ordinary clock-work there can be few possible combinations of trains and functions which have not been tried. The most common variations are, however, the driving of a chime mechanism from the striking train – in fact an extension of striking to quarters rather than actual chiming – and the use of a single spring for alarm and going work in alarm clocks.

In this chapter we are speaking of gears generally, but we shall be considering in particular the gears between the plates. Certain wheels, known as the motion wheels, are as a rule placed between the dial and the frontplate and their rôle is considered more fully in Chapter 4. It is also common practice to place the wheels controlling the sequence of chiming, the ratio wheels, outside the backplate.

Gear Calculation

There is a whole science of gearing which need hardly concern the amateur, especially if he has not the facility to machine wheels and pinions. There are, however, certain principles which it is necessary to understand in general and from a practical point of view when one is faced with the problem of replacing a missing or severely damaged piece. The cardinal point is that of two wheels rolling against each other, where the number of turns made by the larger one will be in inverse proportion to its difference in size; if the wheels are the same diameter, they will both make one turn at the same time, but if one is twice as large as the other, the larger one will only half revolve for one revolution of the smaller. The second point is that wheel teeth are a mechanical contrivance to obviate the fact that rolling wheels wear and slip. The teeth cannot engage unless they are of similar profile on each wheel (or wheel and pinion, the pinion being the smaller and of up to, say, sixteen teeth), and it therefore follows that the numbers of teeth on the two wheels are as closely related to the number of turns made by each wheel as are the wheels' sizes; indeed, the numbers of teeth are merely a different way of referring to the sizes.

In practice, the sizes of wheels are defined in terms of what is called their 'pitch circle'. This is the circle at the point of engagement of the two sets of teeth; it is in fact where the rollers would meet if they were rollers. Obviously the teeth of an engaging wheel and pinion overlap and if we are trying to calculate their true size we must ignore approximately half the depth of a tooth in each case or else, when replacing a wheel, we should find that its teeth merely skated along the top of the pinion rather than meshing with it. Nonetheless, it is necessary to know the true outside measurement of a wheel, including the teeth, and the rule is that the addition of the 'addenda', i.e. the tips of the teeth outside the pitch circle, makes the wheel or pinion the equivalent of three teeth larger than it is at the pitch circle. For example, if a wheel has 60 teeth and the diameter of its pitch circle is 2in, its outside diameter will be bigger by three teeth, standing in relation to the pitch diameter in a ratio of 63:60. Thus, making the calculation $(2/60) \times 63$, we find that the full diameter will be approximately 2.1in.

If a wheel or pinion is missing we shall need to know the number of teeth on the missing item which is necessary to ensure either (according to which wheel or pinion it is) that the centre wheel revolves once in an hour, or that the driving barrel will not make too many turns in the duration and run out prematurely, or that the centre wheel's hourly revolution is correctly changed into a daily revolution for the hour hand. We refer the problem always to a known fact — rotation of a hand on a dial, period of vibration of a pendulum, number of hammer blows in striking, and so on. The trains of antique English clocks, particularly those with seconds pendulums, have been well mapped so that there are tables of standard trains, with the most common combinations of wheels set out, in a number of standard works on clocks and clock repairing. With smaller clocks and modern clocks which are no longer in production, however, the numbers of teeth usually have to be calculated.

In calculating for a missing wheel, one has first to establish the known relevant ends and secondly to work out the desired ratio between them. Possible points of reference are the going barrel or fusee, the centre wheel and the escapement wheel. Since the centre wheel must as a rule revolve once in an hour it is possible to work out the ratio of the train from centre wheel to scapewheel on the basis that each tooth of the scapewheel causes two vibrations in the oscillator. (Where there is a centre seconds hand, the wheel associated with it is normally central and the equivalent of the centre wheel has to be found elsewhere.) The total ratio of the train is found by multiplying together the numbers of teeth on each relevant wheel and dividing that product by the product of the numbers of pinion leaves. Note carefully that it must be *relevant* wheels and pinions; the centre-wheel pinion is not relevant to time-keeping, but to the power-supply and the duration for which the clock will run.

Take, for example, the common case of replacing a platform

escapement, for example on a carriage clock. Here we can discover the vibration of the balance wheel and the number of teeth on the scapewheel from the supplier, but they may not be as before and we may have to caculate the number of leaves which we shall want supplied on the scapewheel pinion. Set the calculation out as a formula, in which the oblique lines represent the gear ratios of the pinions to the wheels. (The contrate wheel is that next to the scapewheel, needed for the scapewheel to be mounted vertically with the platform on the top of these clocks.) If, for instance, the balance wheel vibrated 21,600 times in an hour with a 15-tooth scapewheel, the scapewheel would revolve 720 times in an hour [21,600 ÷ (15 x 2)]. A suitable train might be as follows:

	Centre wheel	Fourth wheel	Contrate-wheel	Scape-wheel
Wheel teeth	80	72	64	(15)
Pinion leaves		8	8	X

Here the fourth wheel revolves 10 times an hour (80 ÷ 8) and the contrate wheel revolves 9 times (72 ÷ 8) for each revolution of the fourth wheel, ie 90 times an hour in all. What you have to do, therefore, is to make this existing ratio up to the required 720:1. Divide the required ratio by the existing ratio (720 ÷ 90) and you have the ratio of the scapewheel pinion (X) to the teeth on the contrate wheel (ie 8:1). As the contrate wheel in this example has 64 teeth, the missing pinion will then need to have 8 leaves to obtain the ratio of 8:1.

As another common example in older clocks, consider the case of the 30-hour clock whose striking is controlled by an outside countwheel (a notched disc with raised sections from 1 to 12, which is considered later in Chapter 6). This countwheel (and the gear attached to it) is quite often missing when one obtains an old movement and it is not a difficult job to cut a new one from sheet metal (see page 143). What is on the face of it more of a problem is to persuade it to revolve at the correct speed. This is of course critical; the locking of the strike at the end of a sequence takes place elsewhere in the train and the train must only be locked between striking, say, 8 and 9 o'clock – it will not do for the strike to stop after giving four blows at 8 o'clock, for it will then be out of sequence for ever after or, as is more probable, will refuse to stop striking at all. The wheel which locks the striking turns once for every blow of the hammer, that is, 78 times in 12 hours (for all these clocks strike simple hours only), and the connection between it and the missing wheel is through the drive-pulley wheel, the great wheel. The total ratio between the missing wheel and the locking or hoop wheel must, therefore, be 78:1. Counting up the number of leaves on the locking-wheel pinion and the number of teeth on the great wheel, see how far you have come towards that ratio – usually locking wheel and great wheel are related 13:1 and therefore the great wheel revolves 6 times in the 12 hours (78 ÷ 13). A further reduction of 6:1 is then required if the countwheel is to revolve only once in 12 hours. The countwheel gear may be driven by a conventional pinion on the end of the drive-pulley arbor or that arbor may be divided into a number of projecting pins, often only four, which constitute a rough pinion. The pinion here will obviously determine how many teeth there should be on the countwheel gear to secure the 6:1 reduction. Often the pinion is missing anyway and you have a choice according to what pinion you can supply or conveniently file up.

The size of any wheel and the number of its teeth are inseparably related, as has been said, and in the case of any one missing wheel or pinion, since the teeth or leaves with which it engages will still be present, it might appear that the circumference of the missing item could be found by multiplying the number of teeth and spaces, once this has been found, by the distance occupied by a tooth and a space. In practice this does not work too well and it is better to use pitch circles as the basis of calculation. The distance between the holes of the missing item and of the wheel or pinion with which it engages is naturally made up of the sum of the radii of the wheel and pinion, and subtracting the known quantity (at its pitch circle) from the total distance will give the pitch radius of the missing factor. From this can be found the pitch circumference (multiplying by 2 and 3.142). Then the 'rule of three' is applied in respect of the addenda of the teeth, the pitch circle in proportion to the full circle being as the actual number of teeth to the actual number of teeth with three added.

Thus in gear calculations in respect of missing wheels and pinions what has to be borne in mind is the interrelatedness of three factors, namely the ratio of revolutions, the ratio of driving to driven teeth, and the ratio of the radii of the known and missing items. The starting point is the known period of revolution of a wheel, such as the centre wheel, or the known frequency of vibrations of pendulum or balance wheel (or, in the case of the synchronous motor, the electric motor's armature as governed by the mains frequency).

For accurate measurement of the depths of gears, and therefore of their sizes, a depth tool is necessary. The depth tool comprises two sets of runners which can be adjusted in their distance from each other, point to point and also longitudinally. Thus it amounts to a pair of infinitely adjustable miniature plates between which a wheel and pinion on arbors can be set up to test for their correct meshing. At the further end of the runners from where the pivots locate are points which can be used as scribers to mark plates accurately with the exact location of two arbors for their best running. These tools, of a size suited to clocks, especially old clocks, are not now easy to obtain, though they can be made. (Instructions for making a depth tool were given in *Model Engineer*, October 1973.) However, for the work of the repairer, rather than of the clockmaker, they are hardly essential. They assist greatly in the test-running of replacement parts, but in the clock for repair we normally know precisely where the wheel has to go. What should, however, be stressed, is that it is essential to do full tests on replacements, examining both how well a wheel and pinion run together by themselves in the plates, and also how they run in the presence of the full train. It is at best a considerable nuisance and at worst a disaster to wind up a reassembled clock when two of its gears do not in fact properly engage.

Repairs to the Gear Train

So far as actual repairs and replacements are concerned, it may be possible to file up a replacement wheel for a larger clock. The required size can be marked out on paper, stuck to the metal with an impact adhesive, or (the traditional manner) the pattern can be scribed on brass which has been rubbed with a moistened rag charged with copper sulphate, which turns the brass black. The cutting of spokes (the 'crossing out') with a fretsaw and filing them off is a long and difficult job, but essential if the replacement wheel is to have a respectable appearance in the train. Motion wheels, however, are not always crossed.

The mounting of wheels on arbors varies according to position (whether or not adjacent to a pinion) and period. Normally a brass collet is driven or soldered onto the arbor. The broad part of the collet serves as a table for the wheel to rest on and the edge is turned over with a punch to rivet the wheel on. It goes without saying that to turn an accurate collet which will run true without a lathe or throw is virtually impossible, although sometimes a collet of suitable size is available in one's parts stock. Exceptionally, if sufficiently thick metal is used and the wheel is accurately drilled, it is possible to drive and sweat a wheel directly onto its arbor, but this is bad practice and obviously unwise with the larger wheels nearest to the spring. The wheels of cheap modern movements are often merely driven onto arbors which have been pressed to raise ridges of metal at the required point. Where the wheel was attached to a pinion and the pinion is still present it is of course practicable to soften the metal with heat, drive the new wheel on, and then to rivet the edge of the pinion down over it.

When a wheel's teeth are worn, even into a slight hollow, on one side the reason is incorrect engagement, either from outset, from a poor repair or from uncorrected worn pivot holes. The holes must be corrected, but the wear still has to be made good or it will worsen and stop the clock. To avoid replacing the wheel, an immediate remedy is to reverse it. Drive the wheel off its collet or pinion with a hollow punch – the riveting metal thrown up in the process will be needed to fasten the wheel back on again. Where a few teeth (up to, say, four) are missing, it is possible to dovetail in replacements filed up for the purpose. Their ends may be hooked at right-angles or flared out, to ensure maximum contact with the solder (Fig 9). In theory, at least, it is also possible to set in similarly a section of another wheel with similar profile. In practice, if the wheel which it is proposed to

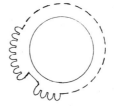

Fig 9

Wheels filed out for insertion of new teeth

sabotage for this purpose is nearly enough identical, it is more satisfactory to substitute the entire wheel, which results in a stronger and tidier job. In barrels and fusees, the best repair is insertion of steel pins filed to shape, as already mentioned, page 31.

The pinion is paradoxical in its apparent simplicity. A seriously worn or broken pinion is one of the more difficult problems for the economically equipped amateur to cope with. Pinions can be filed out of steel, but it is arduous, and there remains the job of true drilling and fixing them to the arbor. Pinion wire – lengths of pinion which were turned and hammered down to the required arbor as needed – is no longer available. Most pinions are in fact solid with their arbors (save for modern brass pinions driven onto steel arbors) and one that is driven on suggests that it is the result of a repair. If disaster strikes, there is some chance of finding a suitable pinion in stock – rather more chance than of finding a suitable wheel – and it may then be possible to adjust the length of the arbor, repivot or hole as necessary and attach the wheel. It may also be possible to soften such a pinion and drill it to fit the arbor. But such convenient circumstances do not seem to arise very often and on the whole the only answer for a lost or broken pinion is to order a replacement, sending the adjacent wheel and the plates to help with the pattern.

For wear rather than breakage there is the possible solution (depending on the value of the clock) of moving either wheel or pinion, or both, in relation to each other, so that a fresh pinion surface is engaged. There are two ways of accomplishing this. First, the engaging wheel can be moved along its arbor. This is done with a hollow punch after application of heat for, whether or not the collet is soldered on, its brass will tend to expand more rapidly than the steel arbor and so will loosen slightly in heat. The alternative is to move the arbors as well as the wheels in the plates. A bush may by mounted in one hole rather proud, and in the opposite hole rather recessed. Such a procedure should only be adopted after consideration, however, for whilst it is not irreversible in that a new wheel and arbor can always be made later, it is obviously a mutilation of what could be a valuable old clock. Moreover, the strain it will stand is limited. Certainly it is inadvisable to try to move a wheel in this way by more than the thickness of metal in the plate, and ideally the portion of bush which is in the plate should be considerably longer than that which is proud of it.

The lantern pinion is a special case and is rather easier to repair than the solid pinion. It consists of two brass 'shrouds' into which are fitted steel pins, known as 'rounds' or 'trundles', and is extensively found on cheaper clock-work. (There is also a variety of lantern pinion, consisting of a single shroud and pins cast in one, open at the other end. This is not amenable to repair.) In the ordinary lantern pinion the shrouds are driven onto the arbor and holes for the trundles are drilled through one shroud and into the other, leaving the second set of holes blind, though sometimes all the holes are taken right through. The trundles are pushed in and the ends covered over with metal by a knurling tool or punch. To replace a trundle, good steel wire of the exact thickness must be used. The remains of the old wire are pushed and pulled out, removing as little metal as possible, and the new wire cut to size and pushed home. If there is insufficient metal to press back over the new trundle, a very little solder can be used. In this operation great care must be taken to make sure that the shrouds remain in alignment. If they become out of true, one must be removed and then sweated back onto the arbor whilst care is taken to prevent the solder from flowing into the trundle holes. If a lantern pinion's trundles are badly worn, it is not good practice merely to rotate them so that a different surface is presented to the wheel. For one thing, they may not stay, and for another, the metal is weakened by the wear. It is more reliable and better craftsmanship to replace the worn trundles, which is not a big job, and it is also better to replace than to try to straighten bent pins.

If a train of gears is to run smoothly and without wastage of power the arbors must run squarely to each other, the wheels mesh at the correct depths, and the pivots meet no resistance. Provided that the pivots are not worn or bent, all this is a matter of adjustment at the holes, a critical area which in practice is adjusted largely by 'feel' and all too often goes by default. The precise amount of 'shake' tolerable between any pivot and its hole is a matter of judgement and experience. Any irregular clatter in the train, when spun round by the fingers, indicates excessive shake somewhere and often, of course, it is possible to see a pivot bouncing from side to side in a hole, moving the oil in the oil-sink with it. From the view of proper engagement, there are also limits to the amount of inward and outer shake which an arbor should be permitted to have between the plates – again, it must not bind, but equally the wheels should engage consistently with the same parts of the pinions. An arbor with too much endshake has from one point of view too much pivot – in effect, a great strain is being placed on the pivot at its point of entry into the plate – and from another it may well have too little pivot, in that these pivots often do not pass right through the plate but lodge half-way through the hole, where serious concealed wear takes place. The root trouble is distorted plates or too short a shoulder on the arbor.

The principal remedy for faults such as these is to 'put in new holes', that is, to bush the holes with new metal. It has, however, to be remembered that the holes are under constant strain in a direction away from the source of power. Consequently, holes which have not been attended to for any considerable period are not round, but oval. The old holes have to be enlarged to take the bushes and so the opportunity is taken to move the centre back to its correct position at the same time so as to ensure proper engagement of the wheels. When remaking a hole, choose a bush which will fit the wheel's pivot, but somewhat closely. Assortments of tapered bushes, and bushing wire which must be tapered and cut to length, are available. Working from the inside of the plate, which is not only neater but also prevents the new hole from falling out, enlarge the hole with a broach until it will

45

very nearly fit the new bush. Work harder on the side which has to be 'drawn' back, so that the bush will be in the exact centre of where the hole should be. The broach will give the hole a slight taper from the inside. The bush is then tapped in until it reaches the required depth. Choose a bush of the right length to allow a projection so that the wheel is moved onto a new part of the pinion if necessary, or to leave a slight recess if endshake has to be added. Ideally, the new hole should pass right through the plate so that it will stand slightly proud and can be riveted tightly back over the old oil-sink on the outside. It is then countersunk to form a new oil-sink. In practice, if the bush is mounted proud inside to move the wheel, this sinking may not be possible, but the bush will still stay firm if it has been well fitted and there is enough metal inside the plate. Broach out the new hole so that it is a good sliding fit for the pivot. Do this with swift light motions of the broach and using oil, so that the hole is hardened and polished inside. Lastly, level off any unnecessary inside projection of the bush above the plate, using the end of a fine file.

If you have a staking set or, better still, a pressing set for bushing plates, all this work will be much simplified, in that you will be able to ensure that the bush ends up completely upright and square in the hole. The job can, however, be done very satisfactorily without these tools provided that time is taken and that thorough adjustments are made, with the wheel and those that engage with it being run together in the plates. If the wheel is at all inclined to run crooked or to stick (assuming that it is not itself bent) the hole must be corrected, probably with another bush, until all is well. This procedure can be carried out if the plates are of good thickness, but if they are very thin the bush may have to be riveted over on both inside and out or, if it is required to stand proud, one may even have to resort to solder. This is not a recommended or a pretty proceeding, but the problem will only arise on a cheap modern movement and there the primary intention is simply to get the clock going again. The quick but short-term alternative is to use a hollow hole-closing punch. It may be mentioned that there is sometimes available a gadget for bushing holes *in situ*, without dismantling the movement. This is not, of course, advisable in a good-quality movement since, apart from anything else, such a method is likely to lead to partial or incorrect diagnosis, and the screwed bushes employed do not produce a good finish, but again it is useful for the cheaper modern movement.

The broken or bent pivot is a serious difficulty without a lathe. If the pivot is broken in a really fine movement, for example a French movement, and in many small modern units, there is little choice but to send it with the plates to have a new pivot or arbor turned. Sometimes, especially if there has been a lot of endshake on the arbor, a pivot retains a long stub which can be used with a proud bush, but it is a risky proceeding and will lead to concealed wear. With the larger and older movements, however, it is often practicable to fit a new pivot. The arbor end needs to be filed flat and softened with heat and then it is drilled to accept a steel wire plug as a tap fit. In doing this as far as possible keep the heat away from the wheel and collet, since they may be soldered to the arbor. The hole drilled should be at least two, and preferably three, times the thickness of the pivot in its depth, and the part of the pivot which is to be driven in may be roughened with a file to provide a better grip. Provided it is not an excuse for poor fitting, a little 'Loctite' solution may be added to seal the job.

The difficulty in all this is of course to drill the arbor perfectly straight and centrally in the absence of a lathe. Sometimes drilling jigs are available for the purpose, though usually only in the smaller sizes. They mount the arbor in a hollow runner, down which the drill is inserted through a guide. It is also possible to rig up an improvement on mere chance by the use of two good quality hand-drills with their chucks facing each other, one being clamped in a vice with the arbor in its chuck, and the other being arranged so that it can be slid along in the same plane. Considerable ingenuity is involved in chucking an arbor firmly in this way when there is a pinion close to the end of the arbor being held. To include this pinion in the chuck is of course to risk breakage. One solution is to drill a brass slug with a hole, tapered as necessary, through the middle, and then to saw it in half in the exact centre. The arbor can then be chucked, assuming there is clearance behind the chuck, round the larger circumference of the slug, which acts as a clamp. Every case is an individual one and it will naturally depend on the circumstances, particularly the time available and the value of the clock, as to whether experiment and ingenuity are to be given scope or the job is to be farmed out. It goes without saying that a pivot replaced by hand must be rounded off and tidied up with a file, preferably burnished, and then very carefully tested between the plates. Where an arbor ends in a pinion and pivot it is preferable to break off the stub of the arbor including the broken pivot and to drill deep into the pinion so that a replacement fraction of arbor complete with pivot can be hammered in with a hollow punch protecting the pivot. Such a section of arbor may well be available from your stock of odd wheels.

A bent pivot can be heated and then, with several slow adjustments, straightened until it will run smoothly in the plates. A sharp twist to the roots of a pivot almost invariably breaks it. It is preferable to bend from farther up with round-nosed pliers, with the pivot protected by a close-fitting brass bush. Where a pivot is already bent acutely at the shoulder there is very little hope of saving it, but sometimes the pivot can be salvaged by using a close-fitting hollow punch placed over it and used gently and progressively as a lever. There is really nothing to lose by trying to straighten pivots, unless they are otherwise defective, because at worst the pivot will break and have to be sent out for replacement anyway.

As regards the nylon or similar wheels and pinions in many modern clocks, these materials are used because they are cheap and easy to produce, do not require lubrication and will suffice, particularly in electric clocks, where there is little driving strain. They are not brittle and it is less a case of replacing a tooth or two than of dealing with general wear and loss of truth. There is really no option but to employ a replacement wheel, from the maker if possible. Sadly, in a way, there is little scope for repair in the latest movements. The amateur may, for his own interest, attempt adjustments, but the fact is that the movements are by design cheap and disposable and the pivots are too fine to repair without specialist equipment. If there is wear in one place, there is likely to be wear almost throughout and the only sensible course is to procure a replacement or substitute movement, which will be cheaper than even the amateur's generous way of costing time spent.

THE GEAR TRAIN

As with mainsprings, so with gear trains, there is a general point to make with regard to lubrication – and it is the reverse point. Do not be tempted to oil the wheels and pinions, though you must, of course, minutely oil all the pivots. It is sometimes supposed that because a train just staggers along dry, it will 'run in' rather better for the application of a little oil. Insofar as minute abrasive particles are concerned, it will run in much better dry. More important, oil will trap the much larger abrasive particles which exist in every household so that the movement's life will be very much shorter than it could have been. The scapewheel is of course lightly oiled, if only via the pallets, but this is the only wheel whose teeth should be 'wet'.

3 ESCAPEMENT, PENDULUM AND BALANCE WHEEL

The escapement is a device for the controlled and regular release of energy, energy used to turn the hands of the clock and also to keep in motion the oscillator which determines how frequently the continuous motion shall be interrupted. The oscillator has a natural frequency of vibration corresponding to such factors as the position of its centre of gravity, the strength of its hairspring, the shape and metal employed (in the case of a tuning fork), and its cellular composition (the quartz crystal). It will always vibrate at the same frequency provided conditions remain consistent.

In fact, of course, such consistency cannot be counted on, for temperature and atmospheric pressure have considerable effect, as in some degree does variation in the sustaining power. Many of these inconsistencies can be avoided, but at a cost and inconvenience which puts the clock (such as the Shortt free pendulum or indeed the quartz crystal clock until recently) right outside the domestic category. Variations in the power of a spring can, however, receive compensation, as has been said, and the other factors by and large do not, as yet, affect timekeeping such as the ordinary household requires. We have also on the end of the telephone, or over radio and television channels, ready access to clocks more accurate than any which could run in our houses. Consequently the pendulum and balance wheel are still reasonable oscillators to control a spring or weight-driven escapement, as they have been for several hundred years. They are also used in many electric clocks where the pendulum or balance is electrically sustained and may mechanically turn the wheels of the train, though this is never a very good arrangement since it interferes with the freedom essential to accurate vibration. Alternatively, the oscillator may close or energise a circuit which advances the train and gives impulse by electro-magnet or by the release of a gravity arm. In such cases there is not strictly speaking an escapement, since there is no energy stored in the train to be released step by step.

An escapement consists of a relatively small scapewheel, with distinctively shaped teeth, and of blades or pins of metal or jewel which interrupt this wheel's revolution and which are called 'pallets'. The pallets are acted upon by the balance wheel or pendulum, most often through the arm on which they are mounted, the crutch or lever. At each vibration the escapement allows a tooth of the wheel to run free, or escape. At the same time the small power of the moving wheel imparts a push, a momentary impulse, to the pendulum or balance. The impulse is given by the carefully sloped surface of pallets or wheel teeth, or both, these being critical angles. Other than these angles the vital matters with escapements are 'depth', ie how close in to the wheel teeth the pallets are set, and 'drop', which is the travel of the next wheel tooth to the next pallet after the previous tooth has just been released. Ideally, immediately one tooth is released, the next one would be locked, but in practice some 'drop', which amounts to a loss of energy, is a necessary freedom. In an escapement all adjustments interact, but the depth largely controls the 'lock', the positive stopping of a wheel tooth by a pallet – obviously the wheel tooth must be near enough into the wheel for it to catch the tooth firmly.

Two further factors are found in many escapements. One is 'recoil', in which the completion of a pendulum's swing after locking makes the pallet continue to move and, by its curved slope, to push the wheel backwards, as can be observed by watching the seconds hand of most longcase clocks. This has the advantages of levelling out variations in the power-supply as they affect the pendulum and reducing its swing after a strong impulse, but it has the drawbacks that it produces friction and reverses the train so that possibly unsuitable pinions are momentarily driving wheels instead of being driven by them. The other factor is 'draw' in a lever escapement. Here the power in the train, by means of the shaped scapewheel and angled pallets, draws the pallet down into the rim of the wheel against the root of a tooth after locking, in order to hold the pallet lever to one side, detached from the balance, which is then quite free to complete its vibration.

Escapements for Pendulums

The Verge Escapement

The verge, probably the oldest of escapements, serves as a good introduction. After perhaps 700 years, it went out of use in the eighteenth and nineteenth centuries for a variety of reasons, particularly the introduction of the long pendulum for longcase clocks – to which the escapement, which involves a large vibration, was unsuited; the need for a contrate wheel in the gearing, when newer escapements did not require this relatively awkward arrangement; and its proneness to wear. It was, however, still used in many nineteenth-century watches, and many of the bracket clocks converted to other escapements in the last 200 years are now being restored to the verge which, when in good condition, gives a very fair performance. Most probably it acquired its reputation for unreliability because of its basic robustness, which led to examples continuing in use long after the time when they should have been serviced and adjusted. Original verges are now very much sought after and those which come the repairer's way usually need a good deal of attention.

The working is shown in Fig 10. Pallets pointing in opposite directions are hung across the diameter of a crownwheel. This wheel has always an uneven number of teeth, usually thirteen or fifteen, and whilst one pallet locks against the vertical face of a tooth, the other pallet is between two teeth on the other side of the crownwheel. When the first tooth is released by the pendulum's swing, the next tooth on the far side of the wheel drops onto the other pallet. The design shown was used from the later part of the eighteenth century. Previously, the pallets were joined directly to the short and light ('bob') pendulum, and there was a knife-edge bearing on the back plate rather than a pivot in a hole. Over the end of the knife edge was fitted a sheetpmetal 'apron' with a block in it, often highly ornamented with engraving, the purpose of which was to prevent the knife from jumping off its vee-block (Fig 11).

The verge, as will be apparent to anyone examining a specimen which has received little attention, is a frictional escapement, ie the pallets are always in contact with the wheel and the pendulum never

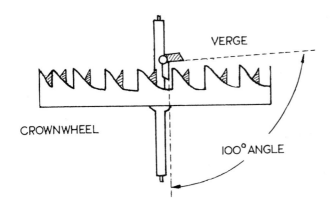

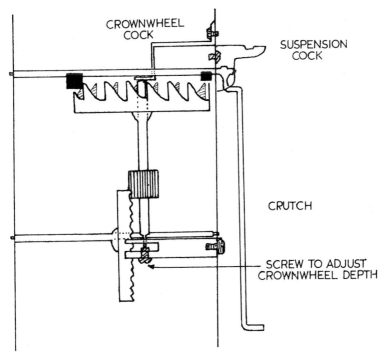

Fig 10

Verge escapement

swings free of the influence of the train. The degree of friction depends on the angle of one pallet to the other, when seen end-on. A large angle results in increased friction and impulse, as well as an enlarged pendulum arc, but also in a tendency of the escapement to lock solid. A small angle was cultivated by later makers since it produced a smaller pendulum vibration, which theoretically will give better timing. The snag is that with a small arc and angle the clock will run sluggishly and will be liable to stop if moved, whereas one of the merits of this escapement, and one of the reasons for its survival in bracket clocks, is its ability to keep going when the clock is moved from room to room. In practice, therefore, a compromise has to be found in the matter of the pallet angle, and the preferred solution is

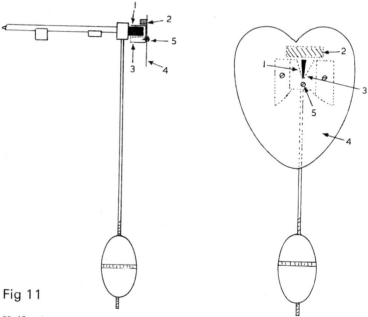

Fig 11

Knife-edge suspension and bob pendulum

1. Knife-edge on pallet arbor
2. Block on reverse of apron which restrains knife-edge from rising
3. Knife-edge bearing
4. Covering apron
5. Screw fixing apron

100° between the pallets, although if a clock defies all attempts to make it come to life enlarging this angle slightly may contribute to a remedy.

The pallet arbor (in fact the 'verge', being named after an ecclesiastical staff of office) is set as close as may be to the crownwheel; the pallets are conventionally a third of the distance between teeth in length and engage with the teeth to two-thirds of their own length, as measured from the centre of the arbor to the tip of the pallet. The pallets themselves may be of solid steel projecting from one side of the arbor (the older sort) or like a flat-ended shovel in shape and projecting on both sides (although of course only one side is used). Their edges on the underside should be sharpened to 45°. Frequently, of course, they are very worn. It is possible to true them up with a file if the wear is not too bad, but generally this will cause a maladjustment of depth which, as they are set so close to the wheel in any case and never on early clocks have adjustable pivot cocks, will be difficult to correct. An effective but rather inelegant solution is to solder pieces of mainspring over them once the wear has been filed or ground out. It may be preferable to make fresh pallets. The traditional method is to make them of thick steel strip, which is twisted to the required angle after being softened by heating to cherry red and slowly cooled – for example, surrounding the whole area in iron wire. The pallet shapes are then cut from the twisted strip and the rest filed round for the arbor. If it cannot be pivoted in a lathe it is possible to drill the ends and to drive in pivot pins, but the chances of coming out with a really true verge this way are not good. It is easier and, in the thinly equipped workshop, likely to produce better results, if one finds a suitable arbor, or makes and pivots one from steel rod, then files two flat slots into it of the right size and angle for the pallets, and uses silver-solder to fix separate pallets into place on the arbor. Adjustments of angle in the pallets, whether old or new, are made by heating the verge to bright red and, using brass-lined pliers and a vice with soft jaws, gently twisting as required. It is advisable to test the surface of the verge with a file to ensure that it is soft before twisting it. Once the change has been made, the whole will have to be tested tooth by tooth in the plates to establish its truth. It can be straightened

55

where needed by peening with a hammer. Finally, the piece is heated again and plunged into cold water to harden.

With regard to the placement of the pallets, there is very little scope for adjusting depth. Make sure that the screw which forms the end stop for the lower pivot of the crownwheel is not loose, and tighten its hole if necessary. The best depth can be set by means of this screw and if there is a large amount of shake in the upper hole it can be bushed. It is essential that the verge fall exactly across the centre, that is the top pivot, of the scapewheel, or the escapement will be impossible to adjust evenly into beat. It may be necessary to move the potence bearing to one side or the other to meet this requirement or, if there is a pivot rather than a knife edge, to move its hole and to bush a new one in the correct place.

The back bearing of the verge escapement sustains a great deal of wear, especially if there is no crutch and the weight of the pendulum on it is added to that of the oscillating verge. If there is a knife-edge bearing it can be sharpened, but this will mean that one side of the verge becomes deeper than the other. It may be possible to raise the potence slightly to overcome this. Alternatively, it will be necessary to replace the knife edge by cutting a slot in its block and fitting a new piece of steel which has been carefully sharpened and hardened. If there is a pivot rather than a knife edge, the hole will almost certainly be elongated and it is necessary for it to be moved and bushed since, once again, the verge must have the same depth on both sides of the crownwheel, and minimal side-shake is required if the teeth are to be locked accurately.

Finally, the crownwheel must be true. If it mislocks on the same tooth on both pallets, either that tooth is 'worn or the wheel is out of truth. It is not within our scope to turn a new crownwheel. The teeth may, however, be generally 'topped' and evened by holding the wheel in the best available chuck and a stake with a hole for the tip pivot to go through, and rotating it against a very fine abrasive. The action must be of the gentlest, for nothing but harm will be done by altering the profile of all the teeth unnecessarily. This profile is of an angle some 25° from the vertical, and the teeth should be slightly flattened at the top rather than very sharp. If individual attention by filing is to be given to some teeth, remember that it is vital that the vertical face of the teeth be left untouched. They have been carefully cut by a division plate and the clock will be lucky to run at all if the distances between teeth are not equal. These faces do not, of course, wear in the action of the escapement and the explanation of any irregularities must be sought elsewhere in teeth or pallets.

The Anchor Escapement

The anchor or recoil escapement generally replaced the verge in bracket clocks towards the end of the eighteenth century, having been in use, in various forms, for about a hundred years, particularly in longcase clocks. It also is a frictional escapement and its resemblance to the verge seen sideways-on will be apparent; it is as if the verge pallets were mounted front-to-back and embracing several teeth, and in fact this was one way of converting a verge escapement (Fig 12). The pendulum is seen swinging from right to left, the left-hand tooth has just been released, the right-hand pallet will fall and the next tooth drop onto it. As the left-hand tooth is pushed round, it slides off the moving pallet, and this action against the curved pallet conveys the impulse to the pendulum. When once the tooth has dropped onto the right-hand pallet, the pendulum will of course keep swinging from right to left, and the pallet will therefore push the right-hand tooth backwards against the power of the train, all the gears being for a moment reversed. This is the recoil.

The anchor escapement exists in a great many forms, some of which are illustrated, for longcase clocks, bracket clocks, and also clocks of various nationalities. The principles are always the same, but the shape and construction of pallets and wheel-teeth differ fairly widely (Fig 13). The family comprises all escapements where the pallets are roughly (sometimes very roughly) in the shape of the flukes of an anchor, where there is recoil and where the pendulum is never free to swing without the influence of the gear train upon it.

The first thing to examine in the working is the depth of locking. Too deep locking will, as with the verge, produce a large arc but at the expense of such friction that the power may well be inadequate to keep the clock going. Too shallow locking will of course result in teeth

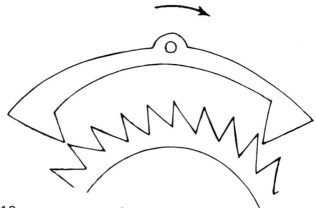

Fig 12

Anchor escapement

Common forms of anchor escapement pallets

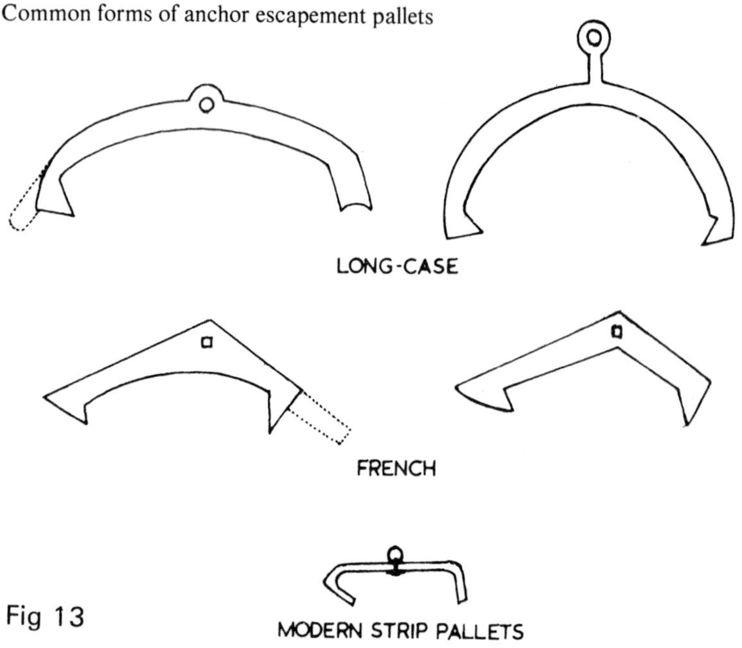

LONG-CASE

FRENCH

Fig 13

MODERN STRIP PALLETS

(*above*) Long-case
(*centre*) French
(*below*) Modern strip pallets

slipping past the pallets or at least in a feeble impulse which may stop the clock. The locking must, in any escapement, be checked on all teeth for both pallets; it should be the same for both pallets and there should be no irregularity caused by a poor tooth. Where the locking is consistently too deep or too shallow, this can be corrected on a modern clock and many continental movements by means of an eccentric screw on the front plate, though the small sideways movement will affect the drops. On the backplate or cock there may also be such a screw. The cock is frequently held in place by screws in slotted holes to make provision for altering the depth at the rear, but in older clocks it will be necessary to move an unadjustable cock. Usually only small movement is required and tapping the cock up or down with the fixing screws loosened may be sufficient. If not, the fixing holes and steady-pin holes must be moved. These are usually well fitted and it may be preferable to move the crutch and pallet pivot hole, which often have to be bushed in any case owing to wear.

The correct drop is the minimum that gives freedom of action. In time one acquires a 'feel' for it – there cannot be too little if the escapement runs, but there may well be too much. If the escapement goes, but sluggishly, the drops are almost certainly excessive and the locking may be too shallow. To find out which drop is at fault is a question of turning the wheel round tooth by tooth and observing the action first on one pallet and then on the other. It is convenient to mark anomalous teeth with a felt-tipped pen. The drops should be the same on both pallets for all teeth. The left-hand drop can be lessened by increasing the depth, but for the right-hand ('exit') drop the pallet face has to be modified.

The usual cause of excessive drop is worn and pitted pallets. If this wear is filed out but the pallets otherwise unaltered, the situation will be no better, except aesthetically, and may well be worse. The alternatives, after trueing, are to close up the pallets, to resurface them, or to move them if wear still remains. Pallets of the strip type are not difficult to close with pliers, though it is as well to check whether they are hard right through or only at the nibs; if a file will not graze the belly of the pallets by the arbor, they will snap if bending is attempted and must first be softened with heat. The solid shiny pallets

59

often found in old French movements are extremely hard and brittle and in any case their shape does not permit of closing them except occasionally when a saw-cut has first been made in the belly. They can be resurfaced with soldered mainspring or it may be possible to move the pallets or the scapewheel on their arbor so that fresh surfaces are engaged. The French pallets are usually mounted on a squared arbor, which is itself fragile. It is wise to file the square down slightly where the pallets have to be moved, before tapping the pallets along the arbor. Many of these pallets are interchangeable, if a little modification is made to the faces after softening, and it is best not to throw out sets which are replaced or spare for any reason. Pallets of a good-quality clock are often screwed to a brass collet on the arbor and this, which may be soldered or friction-tight, can be moved as well as turned so that a new surface of the pivots comes into use.

It is possible, but it is a labour often not worthy of the clock, to make new pallets, either from a detailed drawing of the escapement (the method will be found in technical manuals) or by soldering the old pallets when available onto a suitable steel plate and cutting round them, finishing and polishing the blank afterwards and leaving the nibs bold for final shaping. Basic forgings for longcase and bracket-clock recoil pallets can sometimes be obtained, leaving you the accurate finishing to do yourself. When the work has been done, they need to be hardened and polished, at least on the acting surfaces.

It is as essential here as in the verge escapement that the wheel be true and the teeth even. Often a faulty tooth can be identified in the process of examining depth and drops, and can be corrected individually by filing. Here again the 'inner' edge of the tooth, in this case undercut, must not be touched. Without a lathe it is possible to go some way in trueing a wheel provided that a good drill chuck is available, using the method outlined for 'topping' a verge crownwheel, but lightness of touch is essential since the teeth snap off with great ease if mistreated. Finally, whilst the wheel is usually of thirty teeth and from seven to ten are embraced by the pallets according to whether the clock is a longcase or of a smaller variety, there is a form of this escapement, familiarly called the 'tic-tac', where as few as one tooth may be embraced (see pages 68–9).

Dead-beat escapements

There are several forms of dead-beat escapement, of which the Graham escapement (Fig 14) is regarded as the model. It is a feature of all of them that once a wheel tooth is locked by the **pallets** it is 'dead', there being no possibility of recoil. Their main characteristics reflect this: they are somewhat susceptible to variations of motive power, they employ a very small arc of the pendulum, their power and friction are alike low and, at their best, they are extremely good timekeepers.

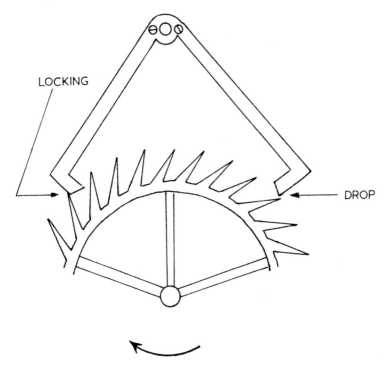

Fig 14

Graham dead-beat escapement

It will be seen that in the Graham escapement the locking face of the pallets is so shaped that, when the pendulum completes its swing after locking, the curve is such as to hold the scapewheel tooth static against the moving pallet. The faces are in fact on a circumference from the pallet arbor. In practice, only the very tip of the tooth comes into action, for the locking is as shallow as it securely can be and that, with pallets of this grade and the jewelled pivot holes which are often fitted, is very shallow. The impulse is delivered by the pressure of the tooth against the foot, the impulse face, of the pallet as it swings away and, slight though the locking is, the tooth must be held on the locking face and not the impulse face or the escapement will merely trip to a stop.

The main test is as usual to check the locking and the drops on all the teeth. The escapement is usually well made and works within fine limits. Therefore all excessive shake in the pivots, whether endshake or sideways shake, must be eliminated before other adjustments are made. The escapement runs on relatively little power and often its pallets are set with real or imitation jewels, so that they rarely become badly worn. In an escapement of this class it will be a mistake to attempt to reface a worn pallet, and in any case it is very difficult since an alteration to one face of a pallet necessarily affects the other. It may be possible to move the pallets or scapewheel, but the real solution is to have new pallets made. An alternative, where a pallet is actually broken, is to saw off the pallet foot, to square the stump, and then to cut half its thickness away. Then, soldering the old end to new steel as a model, cut a new foot from steel. Where this is to be joined to the original pallets a half-joint may be made and screwed up or the new pallet may be tightly pinned in (Fig 15). If the locking is shallow, it is possible to close the pallets after heating. Adjustments for depth are made as for the anchor escapement, but it should be borne in mind that this escapement should be run as shallow as it safely can be and the remedy for weak impulse, provided locking is secure, is to increase the angle of the impulse face rather than to increase the depth.

There are other related forms of escapement used with pendulums. These are the Brocot escapement (pages 65–7), and the escapements found on Vienna regulators (and their imitations), and on 400-day clocks with torsion springs and revolving pendulums. These last two have it in common that their pallets are usually slips of steel slotted and screwed into brass pallet anchors (Fig 16). It will be seen that they differ little in outline, save that the Vienna regulator model has traditional teeth of the Graham type and the 400-day clock has broader, curved teeth. The difference is not functional to the escapement, since the lower part of the teeth is not used, but makes for strength whilst preserving the working part of the Graham tooth. Teeth here and on all dead-beat escapements are extremely fine and fragile and the slight turning over of teeth points is often a contributory cause of trouble if not the main one. The teeth should be checked carefully under a glass. Bends can be corrected by gently gripping in smooth-faced pliers or strong tweezers and pulling – almost stroking – towards the points. Burrs must be lightly filed out, but it is again imperative that the 'inside', near-vertical face of the tooth be untouched, since on this depends the accuracy of the locking.

The steel pallet slips of both these forms of escapement are deceptively simple in appearance and must not be bent – there is indeed no cause to bend them, since they can be adjusted in their mountings which permit small lateral movements as well as pushing in or out. They are cut from a circle whose centre is the pallet arbor and thus correspond to the pallet facings of the Graham escapement. The power at the escapements of both clocks is very small indeed, and these pallets do not wear, but they are often provided with an impulse face at either end so that they can be reversed in the event of losing their edge. (The right-hand and left-hand pallets are different and are not interchangeable.) Both types of clock are provided with eccentric-screw mountings for the pallet staff so that the depth can be adjusted. There is an important difference between them in that the Vienna escapement locks shallow, like the Graham escapement, whereas the 400-day escapement has to lock deeply owing to the large vibration of the pendulum. Here, whilst the wheel is locked and the pendulum continues to revolve, the pallets are banked, at the fork which constitutes the crutch in this mechanism, on the pins or on the movement's frame. The pallet, with its curve, could not in fact reach the rim of the wheel between teeth even without this banking, but it could land on the thick root of the previous tooth. It should ideally be adjusted so that the degree of locking, when the escapement is banked,

Fig 15

Screwed or pinned repairs to dead beat pallets

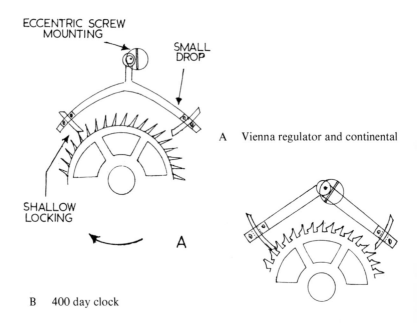

ECCENTRIC SCREW
MOUNTING

SMALL
DROP

SHALLOW
LOCKING

A

A   Vienna regulator and continental

B   400 day clock

B

Fig 16

Forms of dead-beat escapement

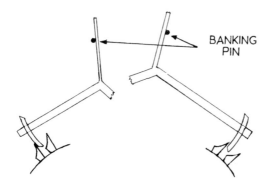

BANKING
PIN

Fig 17

The banking and deep locking of the 400 day escapement

is equal to the distance of the pallet's tip from that root (Fig 17).

The most common fault with both these escapements, apart from bent teeth, is that teeth drop onto one or other impulse face rather than locking. It is best to turn the eccentric screws as a final measure only, for in these escapements adjusting the pallets is not drastic, as in those with fixed pallets where metal taken away cannot be replaced. In this particular connection what is required is to move one pallet into the teeth. It has to be remembered – here as always with escapements – that an adjustment to one pallet produces in effect adjustment to the other and, if one is moved in, the other is likely to need slightly moving out.

The Brocot Escapement

This is the dead-beat escapement so commonly found in French clocks where the escapement is visible in front of the dial. Its proportions are elegant, and synthetic stone pallets give a touch of colour and a suggestion of opulence. The clocks have often received rough handling, are in cases unattractive to modern taste, and are exceedingly dirty, but the escapement is a very good one.

There are a number of variations in the form of the pallets (see for example Fig 18), but these do not seriously modify the action of the escapement, in which the semi-circular pallet pins are as it were a

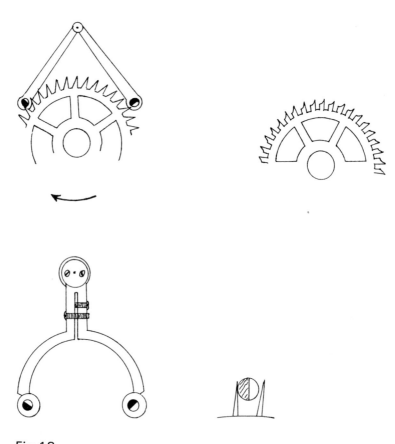

Fig 18

Brocot dead-beat escapement, forms of pallet and wheel teeth

convenient form of the normal Graham shape. They are in fact exactly semi-circular and their size is very critical. They should be halves of a circle slightly less than the distance between two teeth in thickness. It is essential that the flat side of the pallet (which is not used) be radial to the scapewheel and preferable that it exactly face the pallet arbor. There is no distinct locking face on this pallet. The locking takes place on the apex of the semi-circle, and it is vital that it is not even slightly below (and so on the impulse face). These escapements have an eccentric screw mounting for the pallet arbor but, if locking is imperfect or shallow, the arms of the pallets have to be closed, either as outlined for recoil escapements above, or by means of screws provided in the pallet stem below the arbor. The belly of the pallets is normally made with a saw-cut to facilitate this adjustment.

The pallets usually contain ten teeth out of thirty, and the teeth come in two forms, those of the Vienna escapement and those of the 400-day clock escapement. There is no advantage to the latter except possibly strength. The teeth are similarly prone to burring over and must be carefully attended to.

It is a common task to have to replace the pallet pins, or more often one missing pallet pin. These pallets are interchangeable as to side but they vary considerably in size from one clock to another. They may be less than half the space between teeth (though this will result in wastage from excessive drop) but they must on no account be more, or the escapement will neither lock nor give impulse properly. The fixing holes are no guide to the diameter of missing pallets – they were often filled with shellac and some steel pallets were made with thicker ends. The replacement of single pallets leads after a while to an accumulation of a stock of old ones and from time to time one may be brought back into service, for they hardly wear at all. Alternatively, there is no reason why hardened steel pallets should not be fitted to replace 'stone' ones. They must have no taper on them – for the pallet arbor is bound to have a little endshake – and they must represent (by whatever means) round steel rod cut accurately in half lengthwise. Choose rod which is a loose fit between two scapewheel teeth before it is halved. The pallets can if required be made with wide ends for fixing or, depending on the size of the holes, be tapped straight in. If, as was common practice, they are fastened with shellac, it is easy to make subsequent adjustment by softening the shellac with heat – it is useful to countersink the reverse of the fixing hole and to make sure that the shellac flows into it. This helps to secure the pallets.

The 'Tic-tac' Escapement

Reference has already been made (page 60) to the tic-tac as a form of recoil escapement. Its pallets are extremely small in comparison with those of the conventional anchor escapement (Fig 19) and may only embrace one tooth of the scapewheel. They are set close to the scapewheel, and this might be expected to result in a large vibration suitable to small pendulum clocks which were required to be fairly portable. Such would indeed be the effect, were it not nullified by the fact that impulse is only given on one of the pallets. The pallets are usually a solid cam-like block of steel and the entry pallet is of a round shape, being merely for locking and to give a large recoil, the purpose of which is to lessen the vibration and to even out the irregularities of a going barrel. The wheel teeth may be of the Graham shape as is common with pallets which have aspects of both the dead-beat and anchor or recoil escapements, or of the curved pattern used on full recoil escapements.

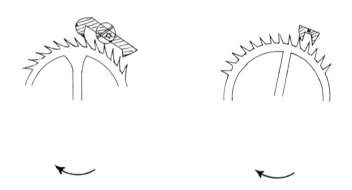

Fig 19

Drum or tic-tac escapements

This is mainly a French escapement, found on many small clocks of the nineteenth century. To the repairer it is one of life's recurrent trials and he would like to be elsewhere when one of these turns up for attention. There are factors, one might say contradictions, in its principle which are very hard to balance up in practice. To add to the problems is the fact that the old silk suspension is common on these clocks since it was favoured for the short and very light pendulum which was used. The suspension does not affect the escapement as such, but it does make the escapement more difficult to set in beat than with the conventional spring suspension.

Obviously, from the profile of the pallets, nothing can be done by bending to close them up. If drop is excessive – and it may well be, since wear and friction are high on this model – resort must be had to refacing the acting surfaces with soldered steel after the wear has been worked out. There is often no adjustment for depth in these clocks, so that if necessary the pallet holes will have to be moved. Correct depth is scarcely a scientific matter here, being what will permit the escapement to run, and best found by experiment; but the pallets are in any case very close to the wheel, there is little room for variation, and if they are set too close they will jam the teeth. In desperation a radical repair may be attempted whereby the rounded pallet is shaped into a straight line, corresponding to the inner acting profile of the other pallet, the extreme point of the old curve being taken up nearer to the horizontal. The change was suggested by Britten in an attempt to impart impulse on both pallets, but if the clock will go without the modification – which certainly makes the jamming of the pendulum less likely – it is best left with its original character. One bright spot is that these pallets are usually fitted with a squared staff and, once the precaution has been taken of ensuring that the square is slightly tapered, they may be gently tapped along the arbor so that fresh surfaces are presented to the scapewheel. These teeth, of whatever shape, are very fine and light, as befits the smallness of the whole, in contrast with the tendency of the escapement to fail to unlock and the attempts of owners to persuade it to run by giving the pendulum 'a good swing'. The teeth are particularly prone to bends and burrs and these must of course be carefully corrected.

Escapements for Balance Wheels

The spring-driven clock is not a modern invention, since it dates probably from the fifteenth century. In its early life the escapement was invariably the verge with a simple unsprung foliot or balance wheel, the latter having at first only one spoke. This combination continued in various forms until the nineteenth century when, as already mentioned, it was still used in pocket watches. There seems, however, to have been a growing distinction before this between the bracket or mantel clock, as a piece of furniture, and the watch. In the late seventeenth and early eighteenth centuries even the smallest clocks tended to be fitted with pendulums, whatever escapement was favoured for them. Meanwhile the truly portable watch continued with the verge and balance. Though there were table clocks – developments of a very long tradition, with balance wheels and horizontal dials – these were in effect large watches. A compromise was reached in the last part of the eighteenth century and after with the popular 'Sedan' clock, which consisted of a small metal or wooden case fitted with a standard verge watch-movement fitted vertically, and in the early nineteenth century the clock with a watch escapement came into its own with the emergence of the 'carriage clock', where again the association with a particular form of travelling vehicle is implicit but debatable. At the same time it became common to mount larger balance-wheel escapements on top of or at the back of clock trains and the era of the modern spring-driven clock had begun, where a balance-wheel escapement, sometimes detachable in the form of a 'platform', is general, the pendulum tending to be used for the more prestigious and immovable pieces. The development has a good deal to do with the evolution of more reliable escapements for use with the balance wheel. Of the verge as fitted with a balance (in clocks) little need be said since, apart from Sedan clocks whose movements are really watch movements, it is only found on valuable antique items and does not appear in the normal course of repair work. It may suffice to note that the principles are as with the pendulum version, but that the verge staff and pallets are mounted vertically and form the staff of the balance. The contrate wheel is used as before to connect the plane of the scapewheel with the different plane of the rest of the train.

The Cylinder Escapement

The cylinder is a frictional escapement closely related to the verge, but with the scapewheel teeth and 'pallets' (ie the cylinder) horizontal. It was thus often known as the 'horizontal' escapement. This escapement became popular on cheaper movements, both watch and clock with platform, in the late eighteenth and nineteenth centuries, though it had been invented a hundred years earlier. It resembles the pendulum tic-tac escapement in embracing one tooth, but it has impulse on both pallets (in this case, both lips of the single cylinder). The cylinder (Fig 20) is part of the balance-wheel staff and it vibrates so as to present alternately an opening and a closed wall to the scapewheel teeth. Thus there are as usual two drops, one as the tooth falls onto the cylinder wall, the other as the same tooth enters the cylinder and drops onto the opposite inside wall. These drops must be equal. The impulse is given by the sliding of the curved edge of the tooth on the lips of the cylinder as it enters, and again as it leaves.

From observation of the action it will be obvious that the sizes of teeth and cylinder are of critical importance. A tooth which is too big will not enter, or will jam inside the cylinder. A cylinder which is too big will jam between two teeth. The cylinder must be the size of the space between two teeth, but with a small allowance for drop. In practice the size of the teeth cannot be altered, for if the teeth are, for example, made smaller to increase the drop within the cylinder, there will then be too large a drop outside the cylinder or, in effect, the cylinder will be to small and the escapement will lack all life. If teeth are corrected, they should be treated with great care, for these tiny steel teeth mounted on vertical stems are brittle and fragile, and they must be altered on the dead face, not on the impulse face or point, since it is here that the division is set.

It is not within our scope to replace or repair a cylinder or balance pivot, and it is doubtful now whether for many clocks the repair would be worthwhile in cost, since most cylinder escapements are fitted to relatively cheap movements which will perform better and be more truly portable with a modern lever platform. (Cylinder escapements in clocks are usually in platforms, and these can be substituted, see pages 82–3.) There are, however, minor adjustments which can be made and it is a pity to change a platform needlessly.

The escapement functions with a small vibration, between a third and half of a revolution of the balance. If it exceeds a half revolution in either direction it becomes locked. To prevent such excess, a fine pin is

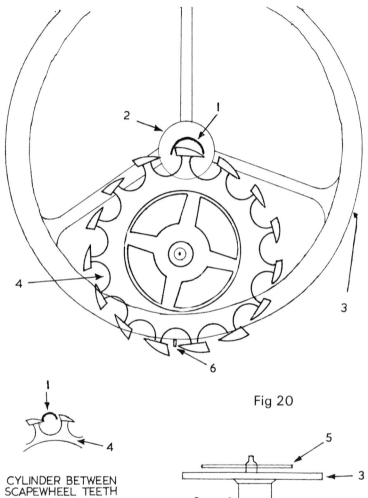

Fig 20

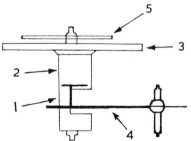

CYLINDER BETWEEN
SCAPEWHEEL TEETH

Cylinder escapement
1. Cylinder
2. Balance staff
3. Balance wheel
4. Scapewheel
5. Hairspring
6. Banking pin on balance wheel

PROFILE OF SCAPEWHEEL
IN CYLINDER

72

fitted to the balance rim and another pin, or a small block, diametrically opposite on the underside of the balance cock. Clearly the wheel must have equal latitude in either direction and therefore if this banking pin is not set dead centrally with the opening in the cylinder, the staff must be held and the wheel turned on it in relation to the cylinder. If the vibration is so large that the pin is hit, it is better not to alter the wheel-teeth's impulse curves, which are difficult to change consistently, but to fit a different hairspring or mainspring (according to the timing of the clock). Where an escapement trips and does not lock firmly within the cylinder it is too shallow. Most cylinder escapements are mounted on a 'chariot' – the balance lower pivot runs in a hole in a movable bottom plate in which is also mounted the cock which supports the upper pivot of the wheel. The chariot is fixed with a single screw and is easily moved, with some adjustment of its steady-pins if necessary. One should, however, ensure that such adjustment is really needed, for the chariots are carefully placed in the first instance and too deep a setting will cause as much trouble as too shallow engagement. Finally, this is a frictional escapement in which the balance is never out of contact with the scapewheel; the lips of the cylinder should receive a little oil, most easily administered by applying a drop on two opposite scapewheel teeth.

The Lever Escapement

The merits of the cylinder escapement are its simplicity and few parts. Its defects are friction and wear, susceptibility to variations of power, the likelihood of a stoppage if it is jolted or moved quickly in the wrong direction owing to the smallness of the balance's vibration, and finally its fragility. These faults were all overcome by the various forms of detached lever escapement which have gradually superseded it.

The lever escapement, as will be seen (Fig 21), conforms to the general pattern of pendulum escapements in having the pallets separate from the oscillator. It has existed in two major forms, of which the 'straight-line' or 'club-tooth' is now universal; a modified version of which is the pin-pallet (which, however, may also be offset), see pages 80–3. We shall concentrate on the action of the club-tooth

73

Lever escapements (old offset rack-tooth and modern straightline club-tooth)

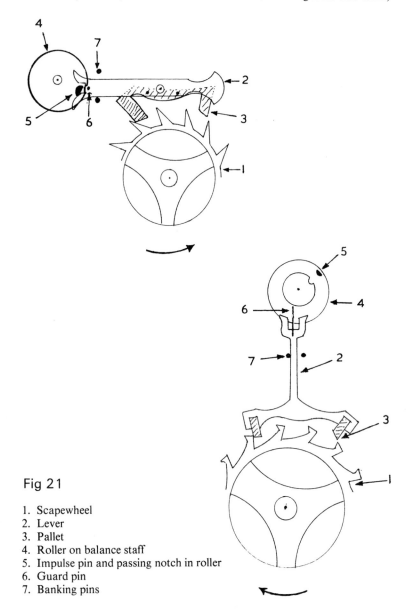

Fig 21

1. Scapewheel
2. Lever
3. Pallet
4. Roller on balance staff
5. Impulse pin and passing notch in roller
6. Guard pin
7. Banking pins

74

form since with an understanding of this and of the recoil and dead-beat pendulum escapements, the working of the earlier offset form of lever will be sufficiently plain. Also good club-tooth platforms are the normal replacement for both cylinder and offset lever escapements in damaged carriage clocks.

The similarity of layout to comparable pendulum escapements is obvious. The pallets are rocked not by a crutch but by the engagement of the 'top' end of the lever with the balance-wheel arbor. This engagement takes various forms and is rather sensitive. Because for a considerable part of its vibration the balance is free of all interference from pallets and train, it must be arranged that the fork at the end of the lever clears the pin on the balance (the impulse pin) at this stage. It is also essential that the lever cannot be 'wrong-sided' by a jolt or movement at this time, otherwise the returning balance wheel would butt against the outside of the lever fork and cause an immediate stoppage. These two ends are achieved by 'draw' on the pallets, by correct banking, and by the design and adjustment of the lever fork and impulse pin.

The semi-circular impulse pin (which may be a jewel) is located vertically in a polished roller on the balance staff. There are two forms of roller, the single and the double (Fig 22). The single roller has a hollow in front of the pin, which points downwards. In the double roller the pin points downwards from the upper blank roller, whilst the recess is in the lower roller. The purpose of the pin is of course to locate with the lever fork and to move the pallets as the balance revolves, the pallets in turn transmitting an impulse from the train through the pin to the balance. The purpose of the roller in either case is to prevent movement of the lever save when pin and fork are engaged. In the case of the double roller a raised pin on the lever, projecting from the fork, obstructs the lever except when this pin, the guard pin, encounters the recess in the roller when the lever alone can cross the balance staff. In the case of the single roller the centre of the fork may be of a pointed triangular form, with the fork below, and this point, again, must coincide with the recess in the roller edge before the lever can pass. Alternatively, an upright guard pin is used.

On either side of the lever are the bankings. They may be pins or

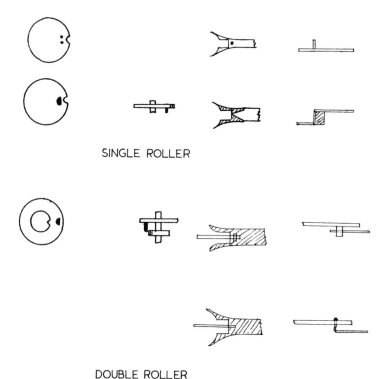

SINGLE ROLLER

DOUBLE ROLLER

Fig 22

Lever escapement, forms of roller and lever
(*above*) single roller
(*below*) double roller

headless screws or merely solid walls surrounding the lever which is sunken into a recess in the plate. Their object is to limit the movement of the lever sideways so that the escapement does not lock to excessive depth. They must also be so placed that, when the impulse pin has passed the lever, the lever fork is in position ready to receive it as the guard is freed by the notch on the roller during the return swing of the balance.

The free operation of this end of the lever is very important. If the lever can swing across the balance when disengaged there is trouble in store, and if the horns of the fork should rub up against the roller the balance will vibrate at best unsatisfactorily. Again, if the lever is short or the roller and impulse pin incorrect, the pin will swing round and land on one of the horns of the fork rather than engaging smoothly. Adjustment to the fork itself is generally ill-advised. What is required is alteration of the banking by bending the pins (twice, to keep them vertical) or scraping metal away from the protruding side walls. This adjustment must be left until the rest of the escapement has been checked. Alternatively, the lever can be slightly lengthened by peening with a hammer or shortened by crimping with pliers. If the lever is of hard steel, it may eventually be necessary to fit a different roller. If the guard pin touches on the roller when the banking has been adjusted and the lever set to length as regards the fork, or if it rubs the back of the roller's hollow, it will have to be shortened. If, on the other hand, it permits too free a movement of the lever, it may be lengthened by tapping through its mounting block, or it may have to be replaced by a longer pin, depending on the type.

The teeth of the club-tooth lever escapement are of a distinctive shape and usually fifteen in number, with pallets containing three teeth and a little space for drop. The wheel is normally of steel (but non-ferrous metal is used in anti-magnetic watches). Observation of the pallet and wheel-teeth surfaces will show that there is slight recoil, that the locking face is at the side and the impulse at the foot of the pallets (Fig 23). The drop can be seen at the second figure. This drop must, as with other escapements, be measured with all teeth and should be the same onto each pallet. The locking, as permitted by the banking and the extension of the pallets, must also be equal on each pallet and should be shallow but secure. The pallet must however be so angled that, after a tooth has locked on its corner, the tooth continues to pull the pallet downwards, forcing the lever into the banking and holding it there whilst the balance is detached and free to continue its vibration. This action and the angle of the teeth are known as 'draw' – the discovery of draw as a means of avoiding displacement of the lever or constant friction on the roller was the main reason for the escapement's becoming popular some fifty years after its invention, without draw, in about 1770. The draw can be tested without the

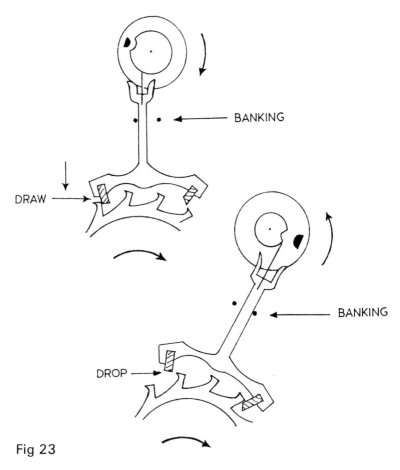

Fig 23

Action of the club-tooth lever escapement

balance wheel by leading the lever until the locking face of a pallet has just engaged with a tooth, when the lever should jump over to the banking if released. The final point concerning the wheel teeth is the 'club'. Their 'top' edges must not be tampered with, since they constitute an impulse face which mates with the face of the pallet. The heel must be sharply pointed – if it is not, there must either be excessive drop or insufficient clearance for the pallets behind the teeth.

If it is decided to move the pallet nibs to correct drop or draw, the procedure is in principle straightforward but in practice tricky until one has the experience to realise how very drastic an alteration this is and how small are the movements required. In modern escapements the nibs are exposed stones set with shellac into the pallet arms of the lever. As noted before, the movement outward of one pallet will normally need to be matched by movement inwards of the other, or the adjustment will be one-sided and far too large. To adjust the stones, the lever should be laid on a brass plate with a hole in it to clear the arbor, and then gently warmed to soften the shellac – gently, because burnt shellac does not adhere. Owing to the shellac fitting, there is normally ample room in a pallet slot for the angle to be altered to increase draw. If further shellac has to be added to secure the pallets, splinters obtained when breaking a stick of shellac will be ample. If the stones are chipped or worn they cannot be repaired. Stones of modern pallets are often interchangeable so that you may have a replacement in stock, but remember that right and left pallets have different impulse angles and cannot be exchanged.

The Lever Pin-Pallet Escapement

The pin-pallet is the cheap version of the old offset lever (though it is also common in straight-line form). Superficially it resembles the Brocot pendulum escapement. Much of what has been said of the lever escapement applies also generally to the pin-pallet, but there are a number of special points.

The scapewheel and lever are of brass, the pallets being steel pins and the wheel teeth having a stubby appearance (Fig 24). The arrangement of fork and impulse pin differs from those on the better types of lever escapements inasmuch as there is normally no roller (though sometimes a single roller is used). Instead, part of the balance staff is flattened to give clearance to the fork, which engages with a steel impulse pin driven into a spoke of the wheel in front of the flat (Fig 25A). The lever fork is of a special shape with two curved horns, and there is no guard pin. The action is basically similar to that of the superior escapements, but in such a pin-pallet the balance staff itself acts as the roller to prevent the lever from crossing at the wrong

Pin-pallet escapement (also made in straight-line form)

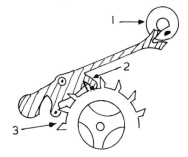

Fig 24

1. Flat on balance staff acts as passing notch
2. Pallet banks on wheel rim
3. Face of tooth undercut to give draw

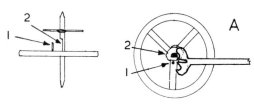

Fig 25A

A   All metal type, impulse pin in balance wheel spoke
1.   Impulse pin
2.   Flat on balance staff

Fig 25B

B   One-piece plastic roller for lever with guard-pin
1.   Tongue (impulse pin) to fit lever fork
2.   Passing recess for guardpin

Parts of modern pin-pallet escapements

moment, since the round parts of the staff fit the recesses in the fork's horns. There is also now in common use a one-piece plastic roller, driven onto the balance staff, and working with a simple fork and guard pin on the lever (Fig 25B). There are often faults in this part of the escapement and time should be spent in ensuring that the clearances are correct and that the lever is fully moved across by the impulse pin. Again, the length of the lever may require adjustment and this can be done by peening or by crimping with the pliers since these levers are brass. The pallet pins are not always dead vertical, and therefore it is necessary to watch that endshake on the lever pivots is kept to a minimum consistent with freedom, or the action will be capricious. Adjustment can be made by bushing the holes in the plate or in the separate top cock which is often used.

There is not usually a separate banking arrangement in pin-pallet escapements, especially in those fitted to clocks. The pallets slide right down the teeth after locking, and bank on the rim of the wheel. This leaves, of course, limited scope for adjustment since the pallets must not bank until the lever fork is clear of the balance staff and the balance can vibrate freely. Banking occurs as a result of draw, which must be strong in this escapement. First test the locking right round the wheel to make sure that when a tooth has just left a pallet the next tooth falls on the other pallet above its centre; ie that the pallet is well round the corner of the tooth and not skating on its impulse face. Shallow locking is a common fault since the pallets are inclined to bend slightly towards the balance wheel through repeated banking on the scapewheel. After locking, the wheel tooth is so shaped backwards that it draws the pallet down into the bed of the scapewheel, thus holding the lever detached from the balance. This draw is tested in the manner outlined above for lever escapements generally. In this instance, however, it cannot be modified by adjusting the pallets. The real remedy is to provide a new scapewheel, but this will only be worthwhile if the escapement is a recent one and wheels are available or still in production. Without replacing the wheel, a little can be done with a fine file to increase the undercut angle of draw on the wheel teeth, but the initial locking surface at the corner of the teeth must not be affected.

The pin pallets should never be bent, since they ought to present a consistently vertical face to the wheel teeth, though sometimes they have never been vertical from the outset. In any case, they tend to snap off if bending is attempted. Adjustments to depth of locking and to the shake of the fork on the balance staff (where again both sides should be equal) are made by bending the brass arms of the pallets slightly as required. Solid brass pallets can be closed after a slot has been sawn into the belly. Broken or worn pallet pins are replaced without difficulty, the old ones being punched out and new ones fitted. They should be of hard steel – sometimes an untapered section of fine sewing-needle will serve. If the old pallets were large the holes may be tightened and thinner pallets be fitted, provided that the whole is readjusted. Thick pallets reduce impulse and the thinner the pallets the better, provided that they have adequate strength, and locking and banking are set correctly for the changed pins.

This escapement needs lubricating and one of the difficulties of the design where there is no separate banking is that oil tends to be dispersed. A drop of oil should be given to the impulse pin, the pallets and the conical pivots of the balance staff as well as to the lever and scapewheel pivots.

Replacing a Platform Escapement

Many domestic clocks upwards of fifty years old were fitted with a cylinder platform unless they were of a specially high grade, and this was normal practice in carriage clocks at the turn of the century and later. These platforms are not of high quality and often reach the repairer with broken cylinder or balance pivots. Except where there are special reasons for preserving authenticity, there is little point in going to the considerable expense of having a cylinder platform repaired. Good replacement platforms of straight-line lever type are available in the standard sizes and they will both perform better and, in many carriage clocks, improve appearance. It is therefore a common task to have to fit a replacement platform. Nonetheless, the wholesale replacement of cylinder platforms, faulty or not, which seems to prevail in some quarters, is to be deprecated. Although replacement lever platforms are of high quality, they have also a distinctive appearance, and in future unnecessary and obvious changing of the original platform of a sound clock could tell against it from an antiquarian viewpoint and must surely offend the craftsman.

Replacement platforms come in sizes more or less interchangeable with the old escapements, provided – and it is a most important proviso – that a scapewheel pinion with the correct number of leaves is ordered. The new platform will have to be dismantled for fitting and it is quite possible at the same time to cut down its plate, smoothing and polishing the sawn edges, to fit any groove into which the old platform may have been placed. It may be possible to use the former mounting holes in the movement plates, or it may be necessary to drill new ones. It is best to order an undrilled platform so that you can place the mounting holes where you find best.

Most scapewheel pinions have eight leaves, but there are some, especially the older ones, with twelve leaves. In a carriage clock a good guide (if you have no platform at all) is the size of the contrate wheel – the 12-leaf pinion engaged with an unusually small contrate wheel, some $\frac{1}{2}$in diameter. Usually you will have the old scapewheel and pinion, but sometimes it may be necessary to make a calculation, as shown in the previous chapter. This will depend on the period of vibration of the balance wheel, of which the supplier will inform you.

The platform will be mounted by four screws on a carriage clock, possibly fewer on the backplate of a mantel clock. You have to place the new platform in position and lightly run it to test, then, having decided its correct position, to mark this on its plate for the mounting holes. They may correspond with the original holes, but they may not, and to proceed by using the old plate as a model and drilling the new one from it is to invite ruining the appearance by having holes in the wrong places. In a mantel clock, the mounting holes often go right through the plate and it is posible to mark the underside of the new platform through the old holes. In a carriage clock, the holes are blind in the thickness of the plate and it is easiest, if they are to be used, to place in each hole a sharp pin or pencil lead so that when the new platform is correctly placed the position of the holes will be marked underneath. Alternatively, the platform can be placed in position and its underside suitably scratched with a graver. If the old holes are plainly unsuitable in position or their threads are worn out, as often occurs, it is obviously best to proceed to drill platform and plates together with new holes, for the old holes will be covered, unless too small a platform is being fitted. Drill the holes in the platform on the large side for the screws. The positioning is very critical in relation to the depth of engagement with the contrate wheel, and the shake in large holes will permit adjustments when the platform is nearly in place. For this reason, large-headed mounting screws and sometimes domed washers are normally used.

The Pendulum and Suspension

Pendulums have been suspended direct from pivots and knife edges, from silk or cord, and from springs. In most domestic clocks they are suspended from a bracket or cock attached to the rear of the movement. This is a convenient but not an ideal arrangement. The pendulum must be as near independent of vibration and disturbance as possible. The arrangement in true regulators and so-called Vienna regulators, whereby it is mounted on a casting fixed to the stout backboard of the case, is as near the ideal as one could hope to come in a domestic clock.

Suspension from a knife-edge (Fig 11) has of course the disadvantage that the edge is prone to wear but, initially, it produces less friction than the set-up with some verge clocks, particularly continental ones, where the suspension is direct from the pallet arbor. Both arrangements have of course long been obsolete. So has suspension from a thread, but only for a century or so, for it is found in nineteenth-century French mantel clocks. Its disadvantages are its inconstancy and the wear on the thread with all but the lightest pendulums. The thread stretches as it is tightened to accelerate the clock, and it affords no support, so that the pendulum is excessively free to swing in any direction against the movement of the crutch (Fig 26).

The spring suspension has been the most common since the eighteenth century but it has taken various forms. The commonest is the simple pinning of the spring between two chops mounted on the cock. The spring may be single or double, as is common in mantel clocks of recent vintage, and it may be fixed rigidly at the lower end to the pendulum rod, or the latter may have a hook hanging over a long pin bushed through the brass chops of the spring. The latter is of course common in clocks likely to be moved, for it protects the spring

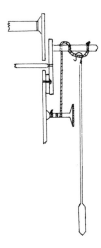

Fig 26

Silk thread suspension

in some degree. The replacement of such springs presents no particular problem, since they are available in many shapes and sizes and in any case are not difficult to make. But it is essential that they *are* replaced if they show the slightest kink or damage because a faulty suspension spring can play havoc with time-keeping as well as lead to irregularity in the escapement and bumps in the night.

Particularly is this rule of replacement applicable to the suspensions of French clocks. These are typically composed of two very narrow springs connected by brass ends, or a single spring split down virtually its whole length. One spring is very easily distorted in relation to another, especially if the pendulum rod's hook is too tight a fit on the pin through the lower end of the spring. (It must, however, be a true fit and not sloppy.) For much of the last century these springs came in an ingenious suspension and regulating device by means of which the clock could be regulated by turning a small square at the top of the dial (usually to the right to accelerate). The arrangement was patented by Brocot in 1840. Both forms (see Fig 27) utilised the lengthening or shortening of the suspension spring – not an ideal practice – as a substitute for altering the length of the pendulum itself for purposes of regulation. In the older form, close-fitting chops were fixed, and the spring raised or lowered between them; in the later form the chops were screwed up or down the fixed spring. Both are good if cleaned out and if the spring acting on the driving wheel is not so tightly

Brocot suspensions (A—earlier type in which spring rises and falls through fixed chops. B—usual later type in which chops rise and fall with spring between)

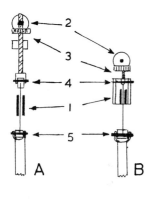
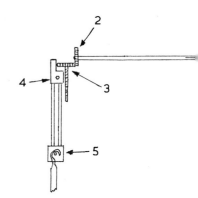

Fig 27

1. Chops
2. Wheel on regulating square from dial
3. Regulating wheel and screw
4. Upper suspension pin
5. Lower suspension pin and pendulum hook

tensioned as to make adjustment from the front impracticable. In the later, commoner, version, a notch is sawn or filed into the top of the brass spring-end in order to give clearance for the driving wheel. In the older sort the spring was pinned into a small saddle attached to the adjusting screw. The pendulums of these clocks also often have rating screws. These should be set roughly to time and the clamping screw turned home to fix the bob, so that one advantage of the system — obviating the chance of wobble in a loose bob — is not lost.

Incidentally, there have been more modern suspensions, particularly American, working on the same system. In these it is usual for the spring to be attached to a brass block running on a fine screw and running through fixed chops.

The corresponding English variation from the simple pattern is that known as the 'rise and fall' suspension (Fig 28). Here a rocking arm is screwed to a bracket on the top edges of the movement plates so that one end of it, which has a right-angled steel pin, projects over the edge of the front plate and the other, slotted to receive the spring, projects over the backplate. The pin on the front plate is engaged by a cam with an indicator hand on the dial, either in the break-arch centrally or to one side. The weight of the pendulum on the other end of the rocking arm holds the pin up against the cam, whilst the suspension spring rises and falls between close-fitting chops, the length of the pendulum thus being effectively altered. The arrangement is excellent, provided that the design is such that there is no slip on the cam. There are many more economically conceived variations, and in these the rocking arm is usually mounted sideways onto the backplate. This is far less satisfactory since the suspension spring is subjected to a constant curvature above the point of suspension.

With the ordinary flat suspension spring, the longer the pendulum, the longer should be the suspension spring. That for a longcase pendulum beating seconds is some 4in long. The thickness and height of the spring control the pendulum and prevent it from yawing, but they also influence its effective length and its arc of vibration, and therefore the clock's time-keeping. Too strong a spring will tend to 'take over' the pendulum's true qualities, whilst too weak a spring will be fragile and inclined to let the pendulum go out of control. It is always wise to replace a spring with one as near to the original as possible. There are, however, occasions when changing the size of spring is a useful way of regulating the clock or of modifying the pendulum's arc. The clock fitted with a thinner spring of the same length will run slow. A thicker spring will accelerate the clock and reduce the pendulum's arc.

There is a special suspension for the 400-day clock to permit adjustment of the pendulum's torsion spring in a circular manner, for

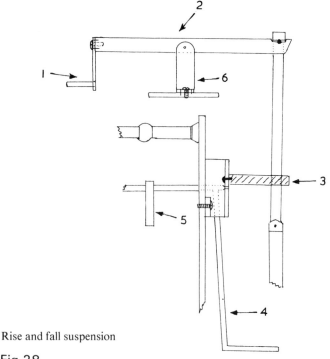

Rise and fall suspension

Fig 28

1. Steel pin engages with cam on regulating index arbor behind dial to cause rocking arm to rise and fall
2. Rocking arm
3. Fixed chops attached to pallet cock
4. Crutch
5. Pallets
6. Bracket attached to top plate across main plates of movement

the escapement is set in beat by this rather than by bending a crutch (Fig 29A). The suspension consists of a narrow spring (which, with the bob, constitutes the whole length of the pendulum, there being no rod) of a certain length, terminating in a hook or the usual chops and pin, to which the heavy bob is attached. The top end of the torsion spring has another brass ending, also screwed on, and it is pinned into a saddle of brass screwed or riveted friction tight to the suspension cock. Regulation is normally by a screw or screws which adjust the effective circumference of the bob by moving small weights in and out. The length and thickness of the suspension spring are extremely critical and, once a spring is on the clock, can only be varied by taking it off and rubbing the spring down with emery or cutting a small piece off. There is of course a rough aesthetic guide to the length from where the bob should hang beneath the movement if it is not to scrape the base or if the gear for arresting it during carriage is to work, but in practice this is little help since a millimetre's difference in the spring's length will greatly affect time-keeping, especially over the long period for which these clocks tend to go unattended. An American guide exists which lists the springs for most models until 1965 but, other than this, the only way is to have a large selection of springs to hand, to use the old one as a pattern if possible, and to keep trying, starting with the thicker sizes and working down. It can be an extraordinarily long and tiresome job. On the other hand, for an old clock, it is one which many professional repairers will not undertake.

There are several variations of this suspension according to the age of the clock. In a common one early in this century (Fig 29B), the saddle is inverted, the suspension pin resting in a notch on top of it, and is clamped adjustably into a block on the backplate. In another form, there is a piece shaped like an 'E', when seen sideways on, of which the middle prong rests in a notch in a small block fitted to the backplate. The lower leg of the 'E' has clamp screws for the suspension spring, whilst the top prong has an adjustable screw which can be lowered into a sink in the block. This arrangement permits movement of the top of the spring backwards and forwards on the suspension, a freedom not allowed in the more standard forms. The block is turned in its adjustable mounting, just as is the usual saddle, to set in beat. It is, however, an awkward device, since the suspension spring cannot be quickly removed for adjustment.

The modern clock differs little from older ones as regards suspension, the principal variation being that the cock is almost always adjustable by screws, usually moving in slots. This is of course for varying the depth of the pallets, whose arbor also rides in the cock, not for changing the length of the pendulum. The pendulum rod is usually hooked over a pin at the end of the suspension spring and this

400-day clock suspensions (A–normal type with saddle riveted or screwed
friction-tight. B–older type with inverted saddle clamped into bracket)

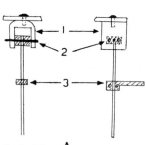

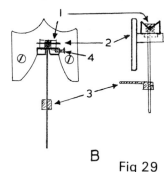

1. Saddle **A**   **B**   Fig 29
2. Suspension pin
3. Forked piece engaging with pallet pin as crutch
4. Screw clamping saddle into bracket

spring may be pinned or screwed between the chops, which are often
merely a split brass rod. The hook may face to back or to front, but it
is more convenient if it goes on from the back since otherwise the
loose spring is apt to be pushed towards the movement and become
inaccessible for hanging the pendulum in the case. Modern clocks as a
rule use shorter and thinner, but broader, suspension springs in
proportion to the total length of the pendulum.

The special regulation of the 400-day clock has already been
mentioned as have those of the rise and fall mechanism and the Brocot
suspension. The arrangements for other clocks present no problems
provided you make sure that a screw with a good thread is used and
that this locates positively at the base, or in some cases the middle, of
the bob. It is obviously essential that the bob is free to move up and
down on the rod over the screw. The exception, lamentably rare in
repairers' hands these days, is the old 'bob' pendulum of verge
movements, where the pear-shaped bob itself is screwed up and down
a fine thread on the rod. With the advent of lenticular bobs of more
aero-dynamic shape this of course became impossible since it was
necessary for the bob always to face in the same direction. Thereafter,
regulation was by the screw beneath, or – in the case of French and
some other bobs – a thumb-screw in the slot of, the bob. English

bracket and mantel clocks often employ devices, sometimes with a thumb-screw, by which the bob is linked to the screw, so that it must, even if rather a close fit on the rod, come down when the screw is turned downwards. In such arrangements the thumb-screw may be mounted on a small bracket above the bob and at the back of the flat pendulum rod. Many, perhaps most, pendulum clocks of this century have flat brass-strip pendulums with a square hole towards the bottom into which the bob and its short section of regulating rod are hooked. This makes for convenience in confined clock cases, but there is the danger that the bob will be a loose fit, swing on its own and become a sort of sub-pendulum stopping the clock. In other instances the bob has a two-pronged hook which passes over a stout pin going from side to side of the rod end and this, provided that the two hooks are level, overcomes the risk of movement in the pendulum itself.

The pendulums of many older French clocks, with their distinctive round bob and slot for the thumb-screw in the middle, are interchangeable and one soon builds up a stock which makes replacement straightforward. It is essential with these also that the double hook to the suspension be a good level fit on the pin passing through the suspension spring. The pendulums of Vienna regulators are, as has been noted, generally suspended in conventional fashion, but through a stout casting screwed to the back of the case. Their rods are usually of elliptical wood which, from the point of view of temperature compensation, is preferable to metal rod, and they have a slot in the middle for the crutch pin.

Few ordinary domestic clocks are fitted with any arrangement to compensate for the variations in pendulum length which changes in temperature bring about. The commonest exception is the continental clock of the last century which is fitted with two glass bulbs of mercury as a bob, the intention being that the mercury expands faster upwards as the whole length of the rod expands downwards, though in fact it is doubtful how far the proportions were strictly calculated. The replacement of these bulbs can be difficult and in many cases, from a horological point of view, will make little difference. Reasonable substitutes can be made of shiny chrome tube, of appropriate thickness and nearly filled with molten lead. The 'gridiron' of alternate brass and steel rods also used in these clocks is usually more decorative than functional, the bob being fixed rigidly to the outside rods so that the different rate of expansion of the inside rods can barely affect it.

From the era of the 'bob' pendulum fitted directly to the end of the verge there has been a good deal of development in the pendulum crutch, the link between the pendulum and the pallets. The simplest – and still one of the most effective forms – is a long wire extension down from the end of the pallet arbor, ending in an angled loop or open-ended slot which is a close, but not a tight or binding, fit on the pendulum rod. For the pendulum of a longcase clock it may be some 5in long, and for a bracket clock with an $8\frac{1}{2}$in pendulum it is usually about 2in long. With this type of crutch, pallets and crutch are removed in one piece from the movement. An unusual form of crutch, more common in turret clocks, is found in French Comtoise clocks, considered in Chapter 6.

There have been many variations in this basic crutch form, in some of which part of the crutch suspends from a cross pin and has to be detached from the pallets before it can be extricated from a square-shaped hole in the pendulum rod. The device of a crutch pin locating in the flat pendulum rod, rather than of a fork embracing the pendulum, is common from the mid-eighteenth century onwards. It was put to good use in Vienna regulators and in many modern clocks in the form of an adjustable crutch, where the crutch pin is mounted freely in a slot at the end of the crutch and can be moved to right or left by the operation of two thumb-screws. The advantage, of course, lies in the ready adjustment of the pallets to the pendulum (for setting in beat) which, with the simple wire crutch, has to be made by bending. More recent forms of adjustable crutch mainly work through using a two-part crutch, one part of which is moved relative to the other on a spring-loaded bearing or catch. There are some where the split suspension pillar achieves the same end by being turned with a screw. These are not entirely satisfactory in that the spring is necessarily bent and there is a tendency for the mounting of the suspension as a whole to be less than rigid. In many modern clocks also there is a pin or lug in the pendulum rod just below the crutch engagement. This is designed to prevent the rod from being pushed up, causing damage to the suspension spring, but it can make rapid dismantling of the pendulum, and crutch assembly, rather difficult.

The replacement of pendulum bobs is not usually difficult. Those of many common types can be bought new, and others are interchangeable, with some adjustment. A handsome large pendulum bob can be made of tubular brass filled with scrap lead, as outlined in Chapter 1, page 23. The lenticular type of bob is made by hollowing a wooden block into a mould for one half of the bob and panning a disc of brass into it. Two discs are made in this way and then sweated together with solder. The rod is lubricated with graphite and then a hole is made in the rear disc and molten lead poured in through a funnel, the whole being smoothed off when set, and the rod being moved during setting to ensure that it remains free. The best weight can only be found by trial and error, although this can be tested with a bundle of scrap on a rod and spring, rather than by making several finished bobs until the right one is found.

The Balance Wheel

The vibration of the balance wheel is determined by the diameter of the balance and by the hairspring in combination. Different balances with the same hairspring will vibrate in different periods, as will the same wheel with a different spring. Although the hairspring's influence corresponds to gravity in returning a pendulum bob to centre, there is no simple or single factor like the length of a pendulum which determines the balance's time of vibration. Moreover, since its vibration is so short, a balance is more seriously influenced by variations of temperature, so that balance wheels, though often not in clocks, are composed of bimetallic strip with open circles, changes of temperature causing the open rims to move inwards or outwards according to the different rates of expansion of the two metals. Nowadays, however, even in the simple balance wheel, modern alloys very much reduce the effect of changing temperature. Finally as to general structure, in the higher grade wheels the period of vibration is controlled not only by varying the length of the hairspring but also by turning out or in large-headed 'quarter' screws on the balance rim.

93

Regulation when running is effected by moving the index lever, through whose curb pins, or a slotted turn-buckle or block, the coil of the hairspring passes. These pins must be close enough together to prevent more than the slightest movement of the spring beyond them and up to its pinning block, but they must also not bind on the spring. If a balance-wheel movement is inclined to gain when the mainspring is fully wound and vibration is large, but to lose later, it is possible to improve the situation by moving the curb pins slightly outwards – they will then have maximum effect in the smaller arcs and accelerate the balance as the mainspring runs down. Broken index pins can be punched out and replaced with new tapered brass pins pushed through from the top and filed smooth there.

So far as poise is concerned, it is necessary to ensure that a balance is properly poised by testing on parallel edges, correcting repeated stopping points by adjusting the screws on the rim or by filing away a little metal. Ideally the balance should also be tested with poise calipers to ensure that it is squarely mounted on its staff, but serious distortion will be apparent if a sight is taken on the level of the rim while the balance is vibrating or if the spring pinning block is nudged by spokes.

Some niceties hardly apply to cheap movements with cylinder platforms or pin-pallets, where the movement itself is not of a quality to give a very minute performance even when fitted with a high-grade escapement. The great majority of domestic clock escapements with balance wheels are in this category. The balances in these escapements (though not necessarily in the replacement platforms) are simple brass rims (sometimes steel in cylinder escapements), uncompensated and unregulated save by the hairspring, though sometimes screws or projections are placed on the rim for effect (when visible) and to distribute the mass to the edge.

One is concerned, in dealing with this class of balance, largely with the more fundamental matter of whether the balance is free to vibrate and to its proper extent. This is governed by the condition of the pivots and hairspring. It is worth attempting to straighten bent pivots (as outlined in Chapter 2, page 47), but success is not often met and, unless the escapement is an unusually good one and otherwise perfect, the cheapest and safest course in the case of a broken pivot is to replace the platform. The same applies to a worn or cracked jewel hole unless one has a jewelling outfit to press a new hole in. (It is occasionally possible, though not recommended, to replace a jewel hole with a fine plain brass bush.) Endstones may, however, be replaced without much difficulty and, as one continues to replace broken cylinder escapements, one gradually collects a supply of spares. The conical pivots commonly found in the screwed bearings of pin-pallet escapements can be cautiously resurfaced if worn. They should be very slightly rounded, not needle-sharp, when looked at under a glass. Naturally, the bearing screws must not be loose in their holes, because the precise shake permissible is within fine limits. Conical pivots need to be well lubricated. The endshake of a cylindrical balance pivot is less easy to determine. One rough method is to press lightly on the top bearing and to see if the wheel slows or stops, in which case the shake is probably insufficient and replacing the endstone may well produce the slight change necessary without your having to risk attacking the pivot. After a while one comes to tell from moving the rim whether endshake or sideshake is excessive or inadequate.

Hairsprings are pinned into open brass collets which are a tight fit on the balance arbor. The other end of the spring is pinned into a brass block or stud which may be pushed hard into the balance cock or be fixed with a screw in one of several ways. The ease of the professional in removing a hairspring is always a little disconcerting. The knack is to spring the collet slightly open with a lever or flat screwdriver and then to slide it up and off the staff almost in one movement. To attempt to prise a spring off from below is to court trouble in the form of a bent staff, distorted spring and cut finger, but there are old springs which will only yield after an initial loosening with a sharp knife – or, better, untapered steel edge – from below, before being afterwards removed in the proper fashion.

Hairsprings can be cleaned with a soft brush and benzine or, in severe cases, by rubbing with pegwood. They must of course be free of oil or other matter which could cause their coils to stick together. It is possible to straighten most bent hairsprings, but often the time is not well spent on a bad case, except when the bend is merely in the outer coil, when it can be straightened by pressing between flat tweezers and then adjusted to a proper curve. A heavily rusted spring must be held beyond redemption. An assortment of springs is worth buying to keep in hand along with those from discarded escapements. Springs from either source do not often turn up both in the right size and with the right collet, and so it is usually necessary to remove the collet and fit a suitable one. This involves cutting out some inner coils in the new spring, bending a flat for pinning, pushing out the pin from the old collet and inserting the new spring. A glass may be useful in the task and it is helpful if the collet can be mounted on a rod in the vice, a projection on this rod stopping the collet from revolving. The very fine pin should have a flat on one side and this must be laid alongside the spring's end as it is pushed in, or the spring will end up distorted beyond correction on its collet. It is easiest to cut the pin to size before finally pushing it home. The last part of the job is carefully to shape the relatively stiff portion of the spring around the collet so that it begins regular spirals as soon as possible but does not touch the collet. The outer coil is bent gradually outwards so that it enters the pinning stud, and the second coil clears both stud and pin.

Mention should be made here of the special arrangement known as the 'floating balance' (Plate 1). This may be found on the back of the movement or between the plates. Usually the balance assembly is on a self-contained bracket and can easily be removed, but the escapement is not of 'platform' type, since lever and scapewheel are mounted between or outside the plates, separately from the balance. The escapement is a conventional offset fifteen-toothed pin-pallet, with the lever parallel to the back plate, but with its fork turned down at right-angles towards the inside of the movement. The balance has a hollow brass arbor, at each end of which is a jewel stone, through which is threaded steel wire stretched tightly between upper and lower balance cocks. The balance must be mounted vertically and thus the bent fork engages with the roller, both being of unusual design. The balance's weight does not rest on the bottom jewel stone (there being no pivot), for a helical hairspring (which in a few years will be a real headache to the repairer needing to replace it) is fixed to the top balance cock and

Plate 1 Floating balance and pin-pallet escapement

descends, enclosing the balance staff, to a collet on the staff just above the wheel. Thus the balance is in fact suspended from its hairspring and revolves round the axial wire merely to give it restraint. The helical spring is wound half way in one direction, then is reversed and wound in the other direction, with the result that as the balance revolves it does not move up and down (as would be the case with a single-directioned spring), since the expansion of one half of the spring is counteracted by the contraction of the other. Regulation is by means of movable weights on the wheel rim. The roller action is unconventional in that the roller is hollow, an open ring of metal, and it is offset to one side of the impulse pin, as is the guard pin, which is in fact an extended fork of the lever bent back. As a result of this (literal) eccentricity, this horn of the lever does not merely enter the equivalent of the roller's notch, but actually goes inside whilst the hoop of metal revolves around it; whilst in the other direction the horn acts more normally on the outside of the 'roller'. There are, as is not unusual with pin-pallet escapements, two parallel impulse pins and a wide slot in the lever fork.

For all this arrangement's interest as a gadget, it is not easy to be persuaded of its substantial advantages. It has not the convenience of a platform escapement of conventional type, though it does avoid using a contrate wheel (as indeed does any platform fitted to the rear of a clock but which normally has the demerit of running balance pivots on their sides). It will not break if jolted, but neither will conical pivots of a normal pin-pallet balance, though the floating balance should wear better. As a substitute for the pendulum on mantel clocks which may be moved around it is not very satisfactory; there are, certainly, not the complications of beat and keeping the clock level which bedevil the average family pendulum clock in the lounge, but nonetheless the floating balance does not respond at all well to changes of position and if seriously out of level will lose or come to a stop through friction of the staff and the steel wire which passes through it. It is obviously desirable in theory to dispense with the constant rotation of a single balance pivot on a single hole or screwed bearing, but it seems rather doubtful whether this alternative in practice offers a great reduction in friction.

So far as servicing is concerned, there are the normal points with regard to depth and locking, somewhat complicated by the unusual roller action. The depth of engagement at the impulse pin is easily altered by bending the lever towards or away from it. There are fixed banking lugs provided so that the pallets bank just short of the scapewheel rim and the lever must have the fullest freedom here, by bending the bankings out if necessary, or the arc will be unsatisfactory. The other thing to ensure is that the balance *does* 'float'. The steel wire must be clean and each jewel hole can receive a drop of the finest watch oil. The position of the hairspring collet on the staff must be such that the hairspring does not draw the wheel constantly onto its upper jewel stone, for if it does so the result will be worse than a conventional pivot, since stone and upper cock are not shaped for running in contact. It must be expected that the balance wheel will vary its vertical position slightly in practice, and it follows that the long impulse pins must be absolutely vertical or performance will vary according to position. Finally, the balance must be properly in beat, and this is effected by turning the hairspring collet on the staff. The lever must not be bent laterally to adjust beat because this will drastically upset the action of impulse pin and roller.

Setting in Beat

Clocks will not run well, and many will not run at all, unless set accurately in beat. Many informal 'repair' calls turn out to be caused not by damage but by moving a pendulum clock to a new location or by a jolt sustained by any type of clock, thus putting it out of beat. The phenomenon is undetected by the owner until the clock stops for want of oil or other cumulative evil, or else is unreliably corrected by putting an envelope under the clock case to level it; but to the clock-lover, hearing a clock which is out of beat is like observing a remediable sickness progress untreated in someone loved.

A clock is in beat when, with the pendulum or balance at the position of rest, the pallets are central, as nearly as possible free of the teeth of the escapement. It will then require an equal movement of the oscillator in either direction to lock or unlock the scapewheel. If the clock is not in beat it will over-lock on one side and barely lock on the other. Its voice will be a limp and uneven 'tick-tock' instead of a sound nearly of 'tock-tock'. More to the point, the pallets or scapewheel may be damaged.

The adjustment of beat in a pendulum clock is more often than not by bending the crutch to compensate for offset pallets or a position of the case which is not level. 'Offset' is of course a relative term, for recoil escapement pallets are often not placed directly over the scapewheel and it is even common for those of cheap mantel clocks to be set at an angle below the wheel. Apart from this, and depending on the size and escapement, many clocks are extremely sensitive to level. Therefore a repaired pendulum clock is best delivered to its owner by the repairer so that it can be set properly in beat where it is to be run. Adjustment can be made from the front, with the hands behind the clock. The rule is to bend the crutch towards the side where the locking is late and weak – the side with the 'tick'. This must of course be done without straining the escapement or pallet pivots, so the crutch is held high up with one hand and bent lower down with the other, a sharp bend being avoided. It can take a good many attempts to find the exact degree of adjustment required. Precisely the same object is achieved by turning the Vienna-type thumb-screws moving a crutch-pin in a pendulum, or by moving the lower part of an adjustable crutch in relation to the higher. Some older French, and many modern crutches, are reputedly 'self-adjusting'. The crutch is screwed friction-tight onto the pallet arbor and the clock can allegedly be set in beat by a hefty swing of the pendulum. Maybe it will be; maybe it will not. One has to be extremely sure that the crutch will move on the arbor before taking this action, and in some cases repeatedly banking the pallets on alternate sides of the wheel will result in an inaccurate setting of beat. There is also the danger of too loose a fitting, when the pallets will never remain in beat even if correctly set. Perhaps it is best to fit this type of crutch fairly firmly and to look upon it as a safety device to be handled much as the ordinary fixed-wire type. It is also possible to set the clock in beat by holding the pallets and moving the crutch on its axis.

The special character of the 400-day or 'anniversary' clock (Plate 2) means that it is set in beat by turning the pendulum suspension cock, which may be riveted or screwed in place. Here the action of the escapement is enormously protracted and the effect of adjustment can be studied at leisure through the hole in the back plate. The simplest guide is that the pendulum will be in beat when its over-swing (that is, the swing after the scapewheel has been locked) is equal with the pendulum revolving in either direction. This is easily observed with the typical modern four-ball pendulum as the position of a ball in relation to a supporting pillar can be noted, and on many of the older types the same can be done with the regulating screws and weights which surmount a disc bob. Since the connection is with a rotating spring, there is no conventional crutch here. The pallets have, coming out of them vertically, an upright pin by which they are moved. A forked piece (usually two brass strips screwed together) engages with this anchor pin, and the other end of this fork is clamped firmly to the flat suspension spring causing the anchor pin to move from side to side as the spring revolves. The precise location of this fork (which amounts to a crutch) and the pin can take a very long time to determine. If the fork is too low the escapement will 'flutter', releasing several wheel teeth for a revolution of the pendulum – it is as if much too long a crutch were fitted. If the fork is too near the suspension, it is as if too short a crutch were fitted and the escapement may not act at all, or it will act without flutter but the pendulum will gradually decrease its arc and eventually stop because the impulse is not getting through to the bob. The correct position has to be found between these two extremes. When it is found, the pallets will move decisively over to the side at each revolution and recoil slightly as a tooth is released. (This recoil is the flexing of the torsion spring, not a genuine recoil from the escapement, which is dead beat.) It should also be said that the fork must neither grip nor act sloppily on the pallet pin. The same principles apply to the 400-day clock with pin-pallet escapement or with wheel teeth shaped other than those in Fig 16.

Balance wheels are akin to pendulums as to beat, but the centering of the balance wheel's place of rest has to be carried out by moving the wheel, and so also the hairspring. The hairspring and where it is pinned of course control the timing of the clock. Therefore, while there may be a little scope for adjustment within the range of the regulating

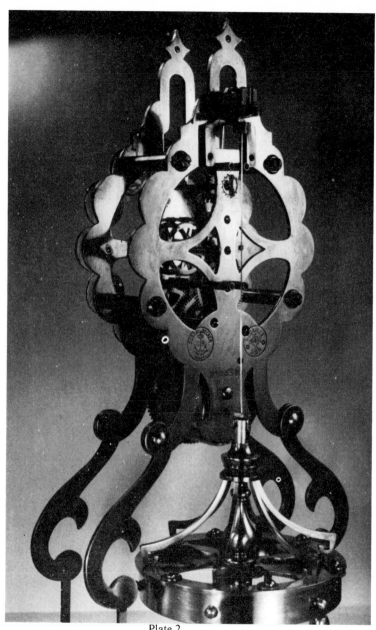

Plate 2
Early 400-day or anniversary clock movement

index (which however should be central after you have set the balance to time and in beat), normally, when the hairspring is moved round, a compensatory adjustment has to be made to its position on the balance staff so that the pin can still be fitted at the same spot although the wheel has been moved round. This is effected, with the balance out of the movement, by slotting a lever or flat screwdriver in the opening of the hairspring collet, springing it slightly open and turning the hairspring the required amount. The balance can then be returned to the plate, the hairspring pinned and the effect of the adjustment observed. The escapement will, if correctly adjusted, be in beat when the impulse pin comes to rest centrally in the lever fork, with the lever in a straight line to the balance staff, or, in the offset type, when the pallets are in appearance as those of a pendulum anchor escapement correctly in beat, or when the banking pin on a cylinder-escapement balance wheel is in a straight line with the centre of the index scale and the scapewheel pivot. It is not, incidentally, always possible to obtain a perfectly even voice from a cylinder escapement since the tooth sounds differently in its drops on the inside and outside of the cylinder, but it is essential that this escapement, with its easily disturbed and rather timid vibration, be set accurately in beat.

4 MOTION AND CALENDAR WORK

Motion work

The motion work comprises wheels, typically between the dial and front plate, but sometimes between the plates, needed to secure a reduction in the centre wheel's hourly revolution to one revolution in twelve hours, and also to operate calendars and ancillary devices. It has been conventional in most clocks for over 250 years for minute and hour hands to be concentric, the minute hand running on or attached to the centre-wheel arbor and passing through a pipe on which the hour hand runs. The second aspect, which we will consider with motion work, is that of the 'clutch' by means of which the clock can be set to time, usually by turning the minute-hand arbor at front or back, but sometimes by turning an arbor geared with it, without forcing all the train wheels round with it and breaking the escapement.

For concentric running, the desired 12:1 gear reduction can only be achieved by means of an offset wheel (the 'minute wheel') driven by a pinion (the 'cannon pinion') on the centre-wheel arbor (Fig 30). The reduction could of course easily be achieved in one step, for example by a pinion of 6 into a wheel of 72. In many older, especially striking clocks, this is in fact the arrangement; the minute wheel and cannon pinion (or 'cannon wheel' as it may then be called) have the same number of teeth and are merely part of the return to concentric arbors, whilst the gear reduction is carried out between the minute pinion and the hour wheel. This had a certain merit in that it was then possible to set off the striking from the minute wheel, the setting off was visible (not hampered by the hour wheel as is setting off from the centre arbor) and the layout allowed more room for the striking mechanism, although it was an arrangement followed sometimes also when there was no striking. More often, particularly in modern clocks, the gearing is divided between all the wheels needed to gain concentric hands. Thus we may find 10 leaves on the cannon pinion driving a minute wheel of 30 teeth, and its minute pinion of 8 driving an hour wheel of 32 teeth.

The minute wheel is mounted freely on a stud to which it is usually pinned, screwed or clipped. Sometimes, however, especially on French clocks, it has a separate overhanging cock which may also serve to keep the hour wheel in place. The general practice is for the hour pipe to fit loosely over the centre arbor or cannon pinion, to which the minute hand is attached; but in older clocks there is often a raised bridge and pipe containing the cannon pinion and arbor and on which the hour wheel runs with its own pipe. In old 30-hour longcase clocks with one hand there is no projecting central arbor as such. The hour wheel may run on a stud or be pivoted between dial and front plate and it is driven in a 3:1 reduction gear from, for example, a 12-leafed pinion on an extended barrel or pulley arbor (Fig 33). In many modern timepieces the minute wheel runs on a stud on the back of the frontplate and the centre arbor and hour pipe protrude through the plate whilst the motion wheels are between the plates (Fig 31).

There are four principal clutch arrangements (Fig 32). In the traditional English form the central pinion is on a pipe and this cannon pinion rides freely on the central arbor but is held against it by a bent-brass spring washer. The washer is tensioned by the forcing back of the cannon pinion when the hand is pinned on against a domed collet. The spring can be bent to tighten it if it has worn flat. Sometimes it has a square hole and there is a square boss to the arbor – this is a good arrangement since it means that when the hand is set to time the washer does not merely turn with it. It is essential with this form of clutch that the washer's ends press against the underside of the cannon pinion, not against the plate, or the clock is liable to stop. In a related set-up common in French clocks, the cannon pinion is friction tight on its arbor, usually with flats and slots on two sides. These pinions are inclined to wear loose after a time, but it is an easy matter to tighten their grip by raising a slight dimple on the inside with a sharp punch, or by lengthening the flats and slots. Pinching of the pipe at the slots is likely to result in a wobbly pinion, however.

Basic front-plate motion work

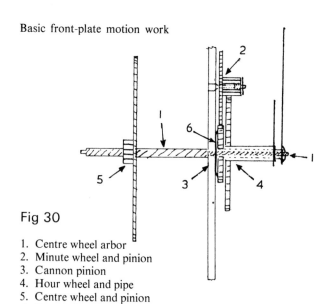

Fig 30

1. Centre wheel arbor
2. Minute wheel and pinion
3. Cannon pinion
4. Hour wheel and pipe
5. Centre wheel and pinion
6. Bow spring friction washer

Basic between-plates motion work

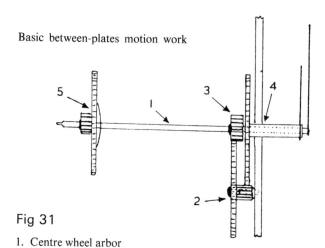

Fig 31

1. Centre wheel arbor
2. Minute wheel and pinion screwed to back of front plate
3. Pinion solid with centre arbor
4. Hour pipe and wheel
5. Centre wheel and pinion held by friction washer

Clutch arrangements

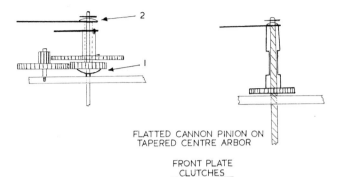

FLATTED CANNON PINION ON
TAPERED CENTRE ARBOR

FRONT PLATE
CLUTCHES

(*top left*)
1. Spring clutch washer
2. Minute hand pinned onto cannon pinion square and thrust by domed
   washer against (1)

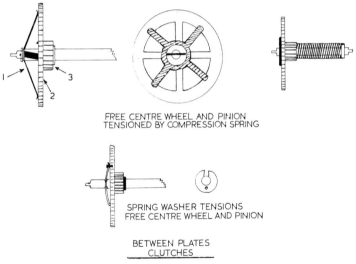

FREE CENTRE WHEEL AND PINION
TENSIONED BY COMPRESSION SPRING

SPRING WASHER TENSIONS
FREE CENTRE WHEEL AND PINION

BETWEEN PLATES
CLUTCHES

Fig 32

(*centre left*)
1. Four-arm spring pinned against centre wheel
2. Centre wheel and pinion free on arbor
3. Washer behind centre pinion

107

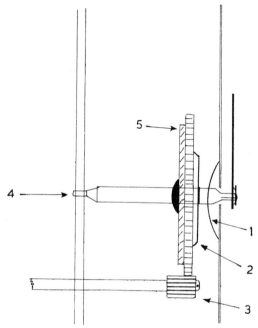

Fig 33

1. Spring washer holds arbor against rear of dial
2. Domed clutch washer springs into slot on arbor and presses hour wheel against seating
3. Pinion on extended arbor of pulley wheel drives hour wheel
4. Hour wheel arbor pivots into front plate
5. Strike lifting wheel fixed to back of hour wheel

The other two clutches are between the plates and are usual on modern clocks. Here the pinion is fixed to its arbor, which is turned from the rear to set to time. The centre wheel and its pinion are free on the arbor but are held tightly up against the arbor boss or a driven collet by either a stiff coiled spring or a 4-armed spring washer. These devices are sometimes pinned together and sometimes driven together at manufacture and they seldom give trouble. In French carriage clocks of late date this arrangement is used with a small washer against the centre wheel. This washer is held in place by a pin or screw which can be removed for cleaning and adjusting. (Earlier carriage clocks used the split cannon pinion frictional clutch.) The essential with these clutches is that their action be firm without binding. This may mean attending to the ends of either type of spring with a file, and bending or stretching the springs to improve tension. Whilst too stiff a spring can damage hand or gearing when time is set, sloppy spring action can produce a perplexingly irregular losing rate in time-keeping. The minute hand is usually pressed or screwed on since, with the fixed pinion on the centre arbor, there is no need to tension the clutch by means of tapered-pin and domed-washer fitting of the hand.

The hour wheel of fusee-driven striking clocks and of single-handed 30-hour clocks is often in two parts and adjustable in itself – for the 30-hour clock this is essential since time can only be set by this hand. The pipe and wheel are kept frictionally together by a round spring washer locating in a slot on the arbor (Fig 33). In a striking clock the arrangement is convenient in that the clock and striking mechanism can be assembled and the hour hand pushed round to match the striking afterwards. Here again, smooth firm action is needed. The washer can be domed or flattened by a tap on the appropriate side from the ball of a hammer.

Simple motion work can cause surprising troubles, although the causes should be plain from observation. One of the most critical actions in setting up an old clock is in pinning on the minute hand with the collet which compresses the sprung washer behind the cannon pinion. One has to have both washer and collet of the right thickness and tension and to choose the right-sized pin. The pin is chosen for fit in the arbor hole, the hand's tightness being adjusted through varying the thickness of the collet or the strength of the clutch washer rather than through choice or fitting of the pin. If all is not well, the minute hand will buckle when the clock is set to time or will slip in the course of running.

Another common difficulty is with the mounting and positioning of the offset minute wheel. If it runs on a stud, this must be firm – sometimes they are riveted, sometimes screwed into the plate – and vertical. The wheels should be checked round with the dial off. These wheels are rarely crossed out and they are inclined to be rough and even burred. It may be that the minute wheel is so positioned that its teeth jam with the cannon pinion or hour wheel. This must be corrected or the clock will certainly stop even though the main movement appears to be in perfect order.

The tightness of springs in the motion work must be attended to. In general, springs may be tighter where the hands are turned by a button or key at the rear. The large 'interwoven' minute hands of later longcase clocks are particularly liable to breakage through trouble in the motion work. The motion work of French clocks is usually of high quality and gives no trouble, but an exception is the between-plates arrangement of later carriage clocks where the screw holding the minute wheel often comes loose and strips the plate thread. The only remedy is to block the hole and re-tap it, or to move the screw and start again.

Special arrangements have to be made in clocks with balances or pendulums of short vibration where concentric seconds hands ('centre seconds') are required. Here the seconds-hand arbor passes through the hollow minute and hour pipes and, because extra wheels are needed in the gearing, the usual centre wheel (once an hour) is itself usually offset to make room for the centre-seconds wheel (once a minute). The displaced 'centre wheel' arbor passes through the frontplate and ends with a pinion which drives the cannon wheel, whose pinion drives the usual 'minute wheel', and thence the hour wheel and pipe (Fig 34). The cannon pinion may be in two parts held together by a spring clip to provide the clutch, in which case the cannon pinion itself can be turned from the front (or by a set-square) but not the cannon wheel. Alternatively, one of the between-plates clutches may be used. The seconds arbor is often very fine and the seconds hand a push-fit onto it.

Arrangement for centre seconds is relatively rare in carriage clocks and greatly increases their value. Two different methods were used. In the first, the older, the rear arbor of the contrate wheel was extended and fitted with a wheel. This meshed with a similar intermediate wheel, and the latter engaged with a centre seconds wheel on an arbor passing right through the movement to the dial, and bearing the seconds hand. These three wheels were on the backplate and ran in a large cock screwed to it. The pinion of the centre seconds arbor was driven by the normal centre wheel. The intermediate wheel was needed

Centre seconds motion work and clutch with displaced centre wheel

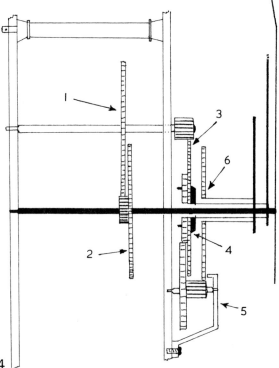

Fig 34

1. Displaced centre wheel, pinion extended through front plate
2. Centre seconds wheel, drives scapewheel pinion (*not shown*)
3. Cannon wheel and pinion linked by friction spring (4)
4. Clutch friction spring links free cannon wheel to fixed cannon pinion on pipe
5. Minute wheel and pinion pivoted in cock which also prevents hour wheel from sliding forward
6. Hour wheel and pipe

to get the centre seconds wheel to run clockwise, but added nothing to the gear ratio. In the alternative arrangement (common in clocks from the Comté, with striking mechanism visible on their backplates) the problem of the distance between the scapewheel and the central dial aperture is solved by using an extended scapewheel arbor, rather than by using extra transmission wheels. The contrate wheel is then set

111

lower than is normal and its arbor is extended to carry the seconds hand. The cannon pinion rides on a pipe fixed to the frontplate and is driven by a wheel (on which operates the clutch washer) running on the extended arbor of the second wheel.

Full regulators dispense with the friction and irregular slack of motion work. Here the centre arbor carries the long counter-balanced minute hand, and the reduction for the hour, which shows on a subsidiary dial, is achieved by wheels between plates. The best Vienna regulators also use such a layout, but many have ordinary front-plate motion work and their seconds hands (which in fact do not register seconds exactly owing to the short pendulums) may be central or subsidiary. The commonest arrangement for subsidiary seconds is of course, as with longcase movements, to plant the hand above the central arbor, fitting it to the scapewheel arbor which is extended for the purpose.

Calendar Work

The calendar in the modern clock is a device of doubtful utility. Since the time when it became common for the ordinary householder to take the daily paper and know the date, the calendar has been making some sort of return, but perhaps rather as a novelty than as a regulator of lives. Its information, which always tended to be conveyed by moving numbers rather than by a pointing index, sorts well with the popular 'digital clock'. Perhaps the main time when it could be of service, namely at the end of the month, is the time when it is least able to help, because the cheap modern clock with calendar has no means of adjusting for the differing lengths of months and offers often either no information at all or merely a disc which the owner turns round manually. But to earlier generations with limited communications even this type of calendar was of importance, and the full perpetual calendar was as much a test of ingenuity and craft for its maker as a practical necessity for the owner, who always had to pay dearly for it.

Calendar mechanisms vary greatly and only some of the commonest are considered here. The repairer meets them, often lacking a wheel or a lever, surprisingly often (though he may not

112

recognise them from what is left) and it is very necessary to know the principles on which they work. Most are driven from the motion work of the going train. There are exceptions driven from the striking work. The latter have the advantage of not imposing the considerable extra friction on the going train, but the disadvantage that the striking must be entirely reliable and that they cannot operate with any strike/silent device.

At its simplest and commonest, the calendar is a day-of-the-month indicator only. It can be set forward by hand for short months. There are two main arrangements in old longcase and bracket clocks (Fig 35). In one, an intermediate wheel is introduced, gearing with a pinion on the hour wheel in the ratio 2:1. The wheel is set on a stud and revolves once in twenty-four hours. A common disposition is for the 'pinion' to have 20 leaves and the intermediate wheel 40 teeth. On this wheel is an eccentric pin which engages daily with a large ratchet ring, mounted on the back of the dial, and advances it daily by one of its 31 teeth. In size the ring is a circle virtually up to the corners of the dial. It is mounted on two rollers towards the bottom and a hook at the top stops it falling back onto the movement. This ratchet is engraved with the numbers 1–31, of which one number shows (the digital clock!) through an aperture below the hands on the dial. The ring is only engaged by the pin for some three hours at midnight – at other times it is free to be adjusted, notably at month ends. If the ratchet is missing a replacement can be cut, after the size has been found from the placing of the rollers and the intermediate wheel's pin. It is advisable to construct a model from card and then to cut round it once it has been stuck to brass. The wheel is best made with 30 teeth, then slit and opened for another tooth to be inserted. In default of engraving, it can be stamped with large metal die-figures, but it is preferable for it to be engraved in the same style as the dial. If, as often happens, the intermediate wheel is missing, its size will be clear from the position of the stud, or the hole where the stud once was, and its count can be calculated simply as 1:2 from the hour-wheel pinion. If the hour wheel clearly never had a pinion then either it or the dial is a replacement, for these pinions are either solid with the hour wheels or fitted with a strong spring washer. This washer comprises another simple clutch,

Simple calendar day-of-the-month indicators

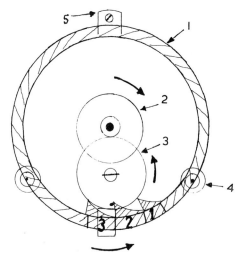

1. Thirty-one toothed ratchet attached to back of dial
2. Hour wheel with pinion
3. Intermediate wheel with eccentric pin
4. Roller on which ratchet rides behind dial
5. Hook to prevent ratchet falling back onto movement

Fig 35

1. Hour wheel with eccentric pin
2. Pin
3. Sixty-two toothed ratchet

enabling the time of day for the calendar date-change to be set.

In the second arrangement, rather later and no doubt cheaper to produce, there is no intermediate wheel and the large ratchet ring is replaced by a disc, again with the numbers 1–31 (half a number at midday) showing through an aperture (or a slot with a fixed arrow pointing to the middle of three days visible). The disc is the front of a ratchet wheel with 62 teeth, operated directly by an eccentric pin on the hour wheel and having a gentle spring click to control its movement. Alternatively, the arbor of this wheel is fitted with a hand moving round a subsidiary dial below the centre arbor. There is no special problem with this set-up save that the shake of the arbor or stud for the wheel tends to become excessive and needs adjustment. With both ratchet devices, but especially with the large ring, there can be trouble with the drive of individual teeth and these should be marked and attended to with a file. Even in the most derelict clock the different forms of aperture, the presence or absence of provision for an intermediate wheel and of hour-wheel pin or pinion indicate unmistakably which type of mechanism should be reconstructed.

A more elaborate calendar will show the name of the month and day of the week. This is carried out by means of a pin on the relevant wheel (ie the preceding wheel in the calendar 'train') which engages with a starwheel and jumper and moves it forward one tooth for each revolution at the end of its own period. The usual old mechanism was that of the 3-arm lever. Here the day of the week, day of the month and month's name are shown from left to right in the 'break arch' of the more imposing dial. There is an intermediate 24-hour wheel with an eccentric pin and driven by the hour wheel, as in the ratchet ring calendar (Fig 36). The pin works in a slot at the foot of the lever causing it to move from side to side every day. One extension of the lever flips round the day-of-the-week lever on one side of the dial and the other turns the 31-toothed starwheel for the day of the month. This day-of-the-month wheel also has a pin, which acts on a pivoted lever to turn the 12-toothed month starwheel at the end of each month. In the short months the clock has, if this is all its provision, to be advanced from the dial. On the return journey of the lever arms, their loose pawl ends, held in place by weak springs, pass by the starwheels without moving them. Again, modern clocks showing day and month usually employ a pin or lever and starwheels. If their display is digital the name or number is in a segment of a wheel or on the side of a cylinder showing through an aperture. The axis of the fully digital clock is at right-angles to the dial so that all the indications can take

Three-arm lever calendar work (*shown from behind*) giving day of week, date of month (simple) and name of month

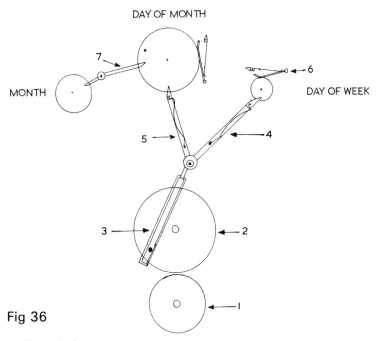

Fig 36

1. Hour wheel
2. Twenty-four hour wheel with eccentric pin
3. Slotted lower arm of lever which engages with eccentric pin and causes three-arm lever to rock
4. Upper arm of lever, with sprung tip, advances day of the week starwheel
5. Upper arm of lever, with sprung tip, advances day of the month starwheel
6. Jumper spring to restrain starwheel
7. Pivoted lever engages with eccentric pin on day of the month wheel and so advances month wheel

this form and with small electric motors (the commonest form of power for these clocks) this is easily arranged.

There are two features commonly found on old calendar clocks which are now largely of antiquarian interest, since our mean time is standardised by radio signals and the 'golden voice', and the state of the moon is rarely of great importance and can be checked from a diary if need be. These features are the 'phases of the moon' and the 'equation of time'. There were various ways of showing the moon's phases, usually involving the emergence of blue or gold moons from behind a crescent or crescents cut in the dial. The commonest was to have two moons emerging alternatively. The mechanism for this at the top of the dial was very similar to that for day-of-the-month indication at the bottom. There was a problem in that no feasible gear train could give an exact representation of the moon's cycle of 29.536 days, but in practice wheels of 59 teeth were used (twice the number), being 60-tooth wheels with a tooth cut out and cramped up. With such a wheel no intermediate 24-hour wheel was needed and the ratchet was operated directly by lever and pin from the hour wheel.

It is extremely complicated to contrive lunar work with continuous action and full gearing, rather than the periodic jerks of the pin and ratchet and, to a lesser degree, the same applies to calendar work. In modern clocks and watches the mechanism still tends to be operated by a pin from the hour wheel or an intermediate 24-hour wheel. This saves gearing and, for the simple calendar which does not discriminate between months, facilitates adjustment, since the mechanism is for much of the time disengaged. Days of the week, months and moonwork can thus be seen as merely extensions of the basic day-of-the-month indication, usually by means of pins and starwheels.

Equation work, though reserved for luxurious clocks, is not a great rarity on clocks of 150 or more years ago when the national standard time did not exist and it was common to check one's watch or clock from a pocket sundial. The equation concerns the relationship of apparent solar time (as shown by the sundial) and the mean time as recorded by a clock. The discrepancy arises because the earth rotates in 23hr 56min, not 24hr as all clocks assume. The 'extra' 4min (actually 3min 55.9sec) is an average over the year, the extremes being with the sun behind the clock by 14min in February and ahead of it by 16min in November. There is agreement only 4 times a year. The mechanism to chart this variation is a cam on the year wheel, on which rests a rack held in contact with it by a spring. The cam, which is called the 'kidney-piece' and resembles the shape of a broad bean, derives its form mathematically from the circle representing exact equation of time, the circle being increased or diminished progressively according to how much the sun is fast or slow by the clock. As the rack is moved up and down by high and low portions of the kidney-piece, it engages a pinion whose arbor carries the hand. The dial normally indicates how much faster or slower the clock is showing than the sun, but the equation is sometimes expressed the other way about.

As anyone who has suffered embarrassment with a cheque as a result of following the hints of a simple calendar watch at the end of the month knows, simple day-of-the-month calendar work is fundamentally defective. However much more practical use it may be, the perpetual calendar has always had the air of completeness for which one has to pay extra money. The full calendar clock taking account of leap years (but not as a rule of the centennial year which is not a leap year) was for long a great luxury and a challenge to the master-craftsman. Indeed, it was more widely available only for a comparatively brief period before it came to be of less practical importance to the late nineteenth-century household. Details differ, but the basic system was established by the eighteenth century. The centre was still the day-of-the-month indication, but connected with the month wheel by gears was a wheel revolving once only in 4 years. Above its teeth, this wheel corresponded somewhat in appearance to a striking countwheel inasmuch as out of a disc were cut 20 slots, representing all the short months over 4 years, the 3 deepest slots being for the 3 Februaries with only 28 days. By means of these slots it is arranged that, according to their depth, the day-of-the-month wheel is advanced an extra 1, 2, or 3 teeth from its last day, so that the first day of the next month is shown correspondingly 'early'. A simpler model has a 1-year wheel and adjusts for the 5 short months, but does not compensate for the longer February in a leap year.

The Brocot perpetual calendar made many of these refinements accessible on mass-produced French clocks from the middle of the nineteenth century, and one does from time to time have one of these clocks in for cleaning and repair. The mechanism (Plate 3) drew on the old methods and its setting up and adjustment are. not difficult provided that these principles are understood. The calendar works off the striking train by means of an intermediate wheel of 24 teeth revolving once in 24 hours. At the other end of this wheel's arbor (ie at the front of the clock) is pinned a disc with a pin near its edge and this pin actuates a lever controlling the calendar movement. Clocks of similar type could be turned out with or without calendars because the calendar is a self-contained unit with a separate bezel mounted below the main dial, and it is driven merely by the daily movement and sprung return of the connecting lever. There is no independent power supply and the working of the lever takes some energy away from the striking train, which in consequence tends to strike slowly at midnight. There are differences between various Brocot calendars but the following remarks relate to a typical example and illustrate the general practice (Fig 37).

The main lever or detent operates simultaneously a day-of-the-week starwheel and a ratchet day-of-the-month wheel with 31 teeth – the arrangement is similar to the old 3-arm mechanism. The fingers of the detent are loose so that they fall back into place after brushing over these teeth on their return by the detent spring. The wheels are restrained by jumpers and springs but the levers are held onto their wheels only by gravity. The day-of-the-week wheel carries a pinion behind the main plate and this is geared into wheels for indicating phases of the moon. This gearing varies according to whether the dial disc is one with 2 or with 3 moons (which are blue) revolving into the round hole in the dial. The day-of-the-month wheel is geared into a 4-year wheel with the countwheel and 20 slots, as mentioned above. The depth of the detent's engagement with the day-of-the-month wheel is controlled by the countwheel on the arbor of the 4-year wheel and a gravity-operated finger. One end of this finger drops into the countwheel slots and the other end is immediately below the lower of the two fingers working on the day-of-the-month wheel. When a short

119

Plate 3 Brocot perpetual calendar movement (differential type)

month occurs, the gravity-lever's tail is in one of the slots and, according to how deeply it falls, the lower finger pushes one of the pins on the day-of-the-month wheel. By its position, this pin ensures that the wheel is advanced an extra 1, 2 or 3 teeth at the end of the month and the normal day-by-day advance by the finger on the ratchet is overridden to this extent. In a 31-day month the gravity-arm is by the countwheel raised above all the pins and the ratchet tooth advances in the normal way by means of the upper detent.

The equation work is the usual kidney-piece if present, but it was more common for the 1-year hand pointing to the names of the

Brocot perpetual calendar (differential type), shown at 29 February, Leap Year, from rear

Fig 37

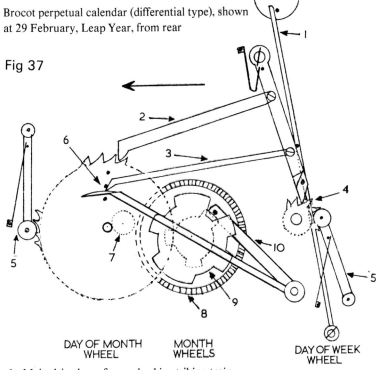

DAY OF MONTH   MONTH   DAY OF WEEK
WHEEL   WHEELS   WHEEL

1. Main drive lever from wheel in striking train
2. Day of month lever
3. Overriding (end of month) day of month lever
4. Day of week starwheel turned by loose tip of lever and carrying pinion below to operate moonwork
5. Jumper and spring
6. End of month pins for (from top) 30th, 29th, and 28th of month
7. Intermediate pinion between day of month wheel and differential wheels
8. Differential wheels, one attached to month countwheel and one to year countwheel (shown dotted)
9. Month and year countwheels. Hand points to month and equation
10. Detent dropping on countwheels and governing fall of overriding day of month lever against end of month pins (6)

months to point also to an outside chapter ring on which the equations during the year were set out. The later version of the calendar (that shown in Fig 37) was in one sense simplified. Here a 1-year countwheel is used, with 5 recesses for the short months, including a very deep one for February. Underneath the 1-year wheel is another wheel, with a different number of teeth and geared differentially to a common pinion mounted on a stud and driven by the day-of-the-month pinion. On this wheel is a cam with 3 projecting lobes. The gearing is such that one of these lobes coincides with the top countwheel's slot for February once every 4 years. As a result, the gravity lever cannot fall fully into the 'February' slot and the day-of-the-month wheel is then advanced, by the second pin, only 2, instead of the usual 3, extra teeth in February. This is of course the only occasion when the second pin down on the day-of-the-month wheel is used.

These calendars give very little trouble, being well-made of substantial metal. The engagement of the pin and driving lever must be carefully adjusted over a period to make sure that the calendar is reliably moved forward each day without unnecessary strain being placed on the striking train in the process. It is essential that the loose ends to fingers move freely, or days will be missed by the failure of the fingers to return to the position needed to turn starwheel and ratchet. The detents must fall freely onto their wheels under the influence of gravity. When setting up in conjunction with the movement, the wheel taking drive from the striking barrel has to be so placed that its pin comes round to the lever at about 11, so that the date is changed at 12 o'clock; it is mounted on a separate cock so that this adjustment can be made when the clock is already assembled. One has also to ensure that this is 12 midnight, and not 12 noon, when starting the clock.

The setting up of the Brocot calendar has to be done with a complete set of circumstances in mind but, once set up, it will never need altering save by moving the main lever – though that will take some time if it is a year or two out in its 4-year sequence. It is easiest to revert to the last day in the calendar when there was a full moon and to place the moon disc accordingly. The countwheel must then be placed so that the slot appropriate to this particular month is engaged, and the day-of-the-month wheel (whose top pin is below the ratchet tooth for the thirtieth day) must be correctly placed in relation to the day of the week. For the differential model it is best to consult an almanac and set up in all detail for the last day of February in the most recent leap year, subsequently advancing the calendar until the present date. Although the day-of-the-week and day-of-the-month wheels are adjustable individually by thumbscrews, they are geared respectively into the lunar work and the countwheel, so that they need to be placed correctly to start with.

5 ALARMS

In a sense the alarm is probably the oldest form of clock; indeed, its bell has given its name to the whole. It originated as a mechanical device to replace repeated manual striking of a bell in civic and religious procedures. For this purpose a wheel corresponding to the crownwheel of a verge escapement was driven by a rope wound round its arbor. Across the crownwheel's teeth would be pallets corresponding to those of a verge escapement and connected to a bell hammer. As the weight revolved the wheel, the pallets were pushed from side to side, swinging the hammer against the bell. From this crude system two developments took place. In the first, gear wheels were introduced, perhaps in the thirteenth century, to prolong the action of the bell. This led, with a foliot or balance replacing the bell hammer, to the clock movement whose principles were to be unchanged for 400 years or more. The other development was to arrange for mechanical setting off of the alarm train, perhaps first by a water clock and subsequently by just such a verge movement.

Alarm mechanism has remained virtually the same, save for the introduction of the electric buzzer or electrically driven alarm, since this inception. It has always been a simple clock escapement in which the scapewheel is driven, usually but not always, by a separate source of power, directly or by very few intermediate wheels, and in which the hammer is directly connected with the pallets. In fact the hammer is very much driven; the rapidity of the alarm having more to do with the power in its spring than with any natural period of oscillation in the hammer considered as a rudimentary pendulum, and the hammer will flutter madly regardless of niceties in its 'escapement'. Essentially, the alarm is thus a subsidiary train held wound and waiting to be let off by the clock's going train. The methods of holding and of setting off have again varied little over the years. For the holding, either the scapewheel itself or the pallets are obstructed by an intruding piece, the other end of which rests on a wheel or cam with a notch, which is turned by the clock and can be independently set. The principle is that of the rocking arm or see-saw. When one end of the pivoted arm drops into the slot, or is forced upwards into it by a spring as the arm's end and notch coincide every twelve or twenty-four hours, the other end of the arm moves and in so doing releases the pallets or scapewheel of the alarm, permitting it to run.

On old bracket (and very occasionally longcase) clocks and on the lantern and wall clocks commonly fitted with alarms, the setting was by means of a dial ring usually concentric with the hand or hands of the clock. This arrangement continued into the twentieth century with cheap hanging clocks such as the so-called 'postman's alarm', a round weight-driven alarm clock which was reputedly used by coachmen to awaken them for their next stage. This setting ring is numbered from 1–12, occasionally from 1–24, and is held friction-tight on the hour-hand arbor. Behind the ring, and solid with it, behind the dial is a thick collet with a single ratchet-shaped notch in which rests one end of the pivoted lever, of which there are various shapes according to where the alarm is placed in the movement. (This is normally at the back, but sometimes at the front, and to one side.) The dial ring is set so that the time for the alarm to go off coincides with the end tail or butt of the hour hand. The ring may have holes to be turned by a peg or have projections for the finger. When the ring is turned, the notched collet behind it is also turned in relation to the hour hand. Then ring and collet rotate with the hour hand until the notch is against the pivoted lever, when the alarm goes off. At this point, owing to the shape of the notch, the ring cannot be moved backwards, but it will travel clockwise raising the lever so that it again interrupts the alarm. In a weight-driven clock there is a separate small weight, together with line and pulley, for the alarm work and in the spring-driven clock a separate spring. Bracket clocks usually have a line wound round the spring's barrel so that the alarm is pull-wound. In the verge escapement alarm, the scapewheel itself was held by means of a pin projecting from the side of the wheel, whilst with the anchor escapement it is normal to intercept either the pallets or the hammer.

125

On these old clocks there is no facility for shutting off the alarm manually when one has heard enough of it, and it often sounds on a sizable and clamant bell mounted on top of the clock with the hammer swinging sideways inside. In a striking clock this may be the same bell as is used for striking.

In the nineteenth century the carriage clock and cheaper developments of it dominated the market for alarm clocks. Most makes of carriage clock could be ordered in various forms with optional embellishments, of which the alarm was one of the more important – indeed many a carriage clock is found today which, though it has no alarm, shows clearly from the proportions of the dial and the space below the hands in the movement that it is one of a range which could include the little sub-dial of an alarm indicator. The alarm mechanism is standard in essentials, using a recoil escapement, a winding square and set-alarm square, and generally a rocking-arm arrangement on the front plate. The alarm wheel is not mounted on the hour arbor but driven, through an intermediate wheel, by the hour wheel in the motion work. The letting off mechanism of the alarm consists of an arbor passing through the plates and ending at the back in the set-alarm square (Fig 38). This arbor is tightly sprung, usually by a coiled spring and washer, or merely a domed washer and pin, against the inside of the back plate, so that it can only move if turned by hand. On its front end, between plate and dial, is mounted a thick collet with the ratchet tooth notch, this time vertically so that the collet has the appearance of being a 1-toothed crownwheel. Freely mounted on this arbor, below the collet, is the alarm wheel, gearing into the motion work so as to turn once in twelve hours – the gearing usually has a nil ratio, its purpose being to get the subsidiary dial below the main one and to ensure that the alarm indicator runs clockwise, but a 24-hour alarm is easily arranged by doubling the ratio. The alarm wheel has a nub or projection on it which on a single revolution rises into the notched collet by spring pressure from behind. There are variations of detail – sometimes the alarm wheel itself has the notched collet as its hub and the set-square arbor ends in a stout pin running into the notch.

The rocking arm is short, going merely to the front plate and being

Alarm mechanism of carriage clocks

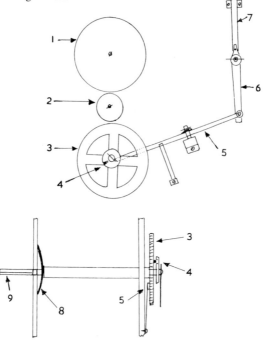

Fig 38

1. Hour wheel
2. Intermediate wheel
3. Alarm wheel free on arbor
4. Raised pin on alarm wheel, and notched alarm cam
5. Rocking arm and spring which presses alarm wheel against cam and also restrains hammer detent
6. Hammer detent, on hammer arbor (oscillates when alarm sounds, released by arm '5')
7. Leaf springs limiting hammer detent movement
8. Washer (or coiled spring) holding alarm arbor friction tight
9. Alarm set-square

pivoted midway in a cock. The end nearest the alarm wheel has a leaf spring beneath it so that the arm presses the alarm wheel up against the collet and the wheel is forced to jump farther up when the projection and notch engage. This movement causes the outer end of the lever to fall towards the plate, thereby releasing a finger attached to the hammer pallets, at which the alarm sounds. The depth of engagement of the rocking arm and the hammer finger is finely controlled by a pointed screw in the lever, and the movement of the hammer when in operation is limited by two leaf springs pressing on it or on pins near its junction with the pallet arbor. The rocking arm is dispensed with where alarm barrel, pallets and scapewheel can be accommodated at the bottom of the plates. Then the finger points up through the plates directly in the path of a rim on the alarm wheel, and is released when the alarm wheel is pushed up by its spring.

These alarms seldom go seriously wrong, but adjustments are often needed to set them back onto the right road. The first concerns the alarm dial. This is frequently a separate ring stuck in a hole in the main dial, and it may be necessary to replace the adhesive (for which 'impact' glue is satisfactory). In this mechanism, of course, the ring should not move, since there is an index, the alarm hand (attached to the collet and setting arbor) to move round the static dial. If the alarm-set arbor becomes loose, the hour wheel turns not only the alarm wheel but also the collet against which it is intended to move, and the alarm hand travels round its dial in company with the main hands of the clock, so that the alarm will only sound at the time shown on the main dial. The cause for this may be too stiff a spring on the rocking lever, but more often it is a set and compressed spring and thin washer behind the setting arbor. The cure is to stretch or replace the spring, or if necessary to fit an additional or thicker washer behind the spring on the arbor.

A second irritating fault is where the alarm will only run in short bursts and needs a jolt to start it again. The trouble appears to be in the escapement, but in fact it is usually in the flat springs limiting the pallet arm's movement. If these are too close and too tightly tensioned they will prevent the pallets from reliably unlocking. If they are too slack, many escapements over-bank and become stuck on one side, and the hammer will jangle unpleasantly on the bell. An associated point to watch is the adjustment of the rocking arm's pointed screw. This must definitively shut off the alarm when the lever's end is up, but allow complete freedom to the pallet arm when it is down. If this state of affairs cannot be obtained, one has to look at the other end of the lever where, for example, the collet and the projection on the alarm wheel may both be worn and have to be stretched a little with a punch, or an adjusting screw at this end of the lever can be moved in or out.

The springs on these alarms are weak and easily broken. The clickwork also is fragile and will permit the determined owner to wind the alarm backwards with serious result, although sometimes the inner end of the spring is bent rather than broken. Attending to a broken alarm spring, the result of over-enthusiasm in winding, is a common repair. In these circumstances the spring usually breaks at the outer end and there is little difficulty in making a new hole for the barrel hook rather than in unnecessarily replacing the spring. Drastic but simple action is similarly needed if the alarm has been set backwards, thus breaking notch or nib. Another fault, obvious enough but easily overlooked when these alarms go awry, is the erratic action of the whole or the seizure of the hammer owing to worn pivot holes for the alarm pallets. They are under constant pressure from a direct spring and after a while the hammer and pallets cease to run square to the scapewheel. These holes should always be checked and bushed if there is any doubt about them.

Whatever its external appearance and the minutiae of its workings, the modern alarm is based quite clearly on its forerunners. The alarm is still a low-geared escapement and the setting and release are still performed, in general terms, by the coincidence of a manually adjustable notched piece with a 12- or 24-hour wheel which jumps into place to set off the alarm. Sometimes the alarm is, as usual in a carriage clock, offset, and there is a subsidiary dial. More often now, the alarm wheel is geared to a central pinion with a broad and short pipe to which is fitted the sweep alarm hand. This should make it possible to set the alarm more accurately, but frequently in fact there is play in the wheels and the alarm hand is so short in relation to the chapter ring of the dial that there is no advantage. Where there is a subsidiary dial a common arrangement is to have a set-square arbor and pin mounted over a sprung alarm wheel carrying the notched cam. The spring beneath the alarm wheel can take the form of bent steel strip, a projection of which leads down between the plates and obstructs the hammer tail – this use of the spring to perform the work of the rocking arm in the older systems is very characteristic of modern alarms. A common arrangement for a central alarm hand is to have the notch in the underside of a many-toothed wheel which engages with a pinion on the set-square for setting purposes. (This is a good arrangement in that the gearing restricts play in the setting of the alarm.) Beneath this notched wheel is the alarm wheel proper, with a projection on it, pressed up by a spring and ready to jump into the slot above. The tensioning spring may continue right across the clock, its extremity being used as the stop on the hammer tail. Whatever contrivance is used, it is as essential as in the carriage clock to ensure that the release, when it occurs, is total. If the release detent is knocked by the vibrating hammer the alarm may run intermittently or be damaged or, if the engagement is too slight, may go off at irregular times.

The silencing mechanism is a distinctive feature of the modern alarm as it emerges from the era of the carriage clock and the bracket clock. The arrangements vary somewhat in detail but are clear enough from observation. There is usually a button or slide, at the top or back of the case, which interposes a stop either on a scapewheel tooth or by pressing onto the pallets or hammer. There may be a separate spring to keep the stop in place. More often in modern clocks, its engaging surface is so shaped that it both stays down by force of the pallets or teeth against it, and jumps out with the help of a small spring when the train is reversed by winding the alarm. Alternatively the spring's clickwork may release the stop. In that case, if you do not wish to have the benefit of the alarm daily, you have in an 8-day clock the option not to bother to wind it up, even if there is only one spring since it is the act of winding which releases the alarm silencer. In a single-spring clock of 30-hour duration, however, you do not have this option, and here the silencer is usually of the type which stays on, or can be overriden by applying a small catch. Sometimes there is a 'repeating' or intermittent sounding device, familiarly known as a 'snooze'. This is arranged by having a pawl dropping onto a large-toothed ratchet on the arbor of a wheel early in the going train. (The spring of this pawl is too weak to significantly obstruct the going.) When the pawl is engaged in a ratchet tooth it stands in the way of a projection on the hammer and restricts it from moving enough to trip on its escapement. When the ratchet moves on as the going train revolves, the pawl rises onto a tooth of the ratchet, leaving the hammer free to sound until the ratchet tooth is passed and the hammer is again restricted. When the hammer is obstructed by the normal set-off arrangement, the pawl merely rises and falls idly on the ratchet.

The use of a single mainspring for alarm and going trains makes the movement considerably more compact and may have the advantage (depending on one's habits) that it is impossible not to wind the alarm. For this purpose a gearwheel is fitted to the end of the arbor where the clickwork would normally be and the click is fixed to this large wheel (a sort of toothed and revolving barrel top), turning as the alarm goes off (Fig 39). Sometimes the back of the wheel has raised ratchet teeth round it and these engage with a domed spring washer with several arms on the squared arbor, and this acts as the click (Fig 6). In effect, therefore, we have a wheel attached to each end of the mainspring (the second wheel being the teeth of the going barrel) and if one is wound up the force of the spring tends to turn both ends. Energy cannot be released through the alarm train, save for the time (which has to be severely limited) when the alarm is actually running. A simple stopwork is fitted which prevents the alarm from running for more than one turn of the spring, so that there is always ample power left to take the going train round through its duration even if the alarm is allowed to run twice in one day. This stopwork acts on a spring at one end of its cycle so that it can be overridden to permit several revolutions in fully winding. Usually the clickwork of these clocks cannot be disengaged and the mainspring has to be let down with the escapement removed.

There are variations in design, particularly in the click and stopwork, but generally this ingenious arrangement presents no special problems until the mainspring breaks. Then, as the top wheel may be riveted to its arbor just above the barrel, it is no easy job to remove the spring or, having removed it, to refasten the pieces together strongly enough afterwards. In fact, so low is the price of such clocks now, that it is hardly worth contemplating such a repair if time and labour are to be costed and charged. If the movement is still in production it may be possible to obtain a replacement barrel, but

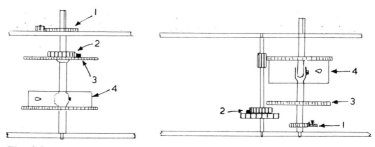

Fig 39

Going and alarm trains driven from one spring
(*left*)
1. Simple stopwork allows alarm only one revolution but by-passes to fully wind
2. Wheel driving alarm train, free on arbor but connected by clickwork
3. Wheel driving alarm train, linked by clickwork
4. Going barrel drives going train
(*right*)
1. Simple stopwork by-passed for winding
2. Alarm hammer-wheel free on arbor but connected by clickwork
3. Wheel driving going train connected by split arbor to inner end of mainspring
4. Going barrel drives alarm train

often parts are unavailable and the owner and repairer alike are invited to send the lot back to the maker. It is worth noting that the parts of modern alarm clocks, outwardly simple and similar, are rarely interchangeable and do not have a sufficient body of material in them to enable them to be adapted. Further, a number of the wheels which would normally require long-term lubrication and take no great strain, are made of synthetic materials so that lubrication is not needed. Such parts cannot be worked upon as could their former metal equivalents. In practice the repair of small alarms tends to be limited to removing the movement from its case and taking the dial off to make small adjustments to the pin-pallet escapement and to the alarm-release detent and spring. One will go further for a friend or out of interest, but there are limits to what can be done and it is rarely economical.

Variations in alarm mechanism are legion, but if the common points are known there should be no great difficulty in dealing with any alarm clock that comes in for attention. There are two further points which often crop up. The first stems from the shape of the notched

132

piece. This must be ratchet-shaped — it must have a vertical side so that the alarm is suddenly released and does not gradually slide into action, but also it must have a sloped side so that the alarm wheel's projection can turn away from the notch under the normal action of the clock. As a result, no alarm clock can be set backwards. Many repairs stem from clumsy attempts to turn the setting hand or square in the wrong direction. They include bent arbors and broken projections on the alarm wheel, but they are not too difficult to make good provided that the wheels are metal. Sticking a new nub, or trying to chisel a recess, for a nylon alarm wheel is less satisfactory. In the better-class clock the alarm-setting mechanism is often fitted with a stout click to prevent this sort of thing from happening.

The second point is that of setting up or correcting a loose alarm hand. These hands are normally press-fits on their arbors and they must be tight and not touch on dial or glass. The procedure in fitting them is to set the going train to the position where the alarm is released, but no farther. The alarm hand is pressed on firmly at the time shown by the main dial. It is then wise to take the main hands round twelve or twenty-four hours checking that the alarm does release again exactly at the time shown. Make sure that the dial is firmly in place when the alarm is being set up, for there is usually enough play in the wheels without introducing the complication of a moving dial, and there are some, but not many, things more maddening in clock-work than assembling and casing a movement only to find that the alarm does not go off at the time which the hand suggests. It is also wise, before returning the movement to its case, to check that there is no movement in the alarm hand as the main hands are turned. Such movement, as has been said, indicates a slack spring on the arbor of the notched cam, the set-alarm arbor.

6 STRIKING AND CHIMING

Striking

When the dial is removed, the apparent tangle of wheels and levers in a striking or chiming clock can be rather daunting, especially if the clock is a compact modern one with unfamiliar detail. It is necessary to bear in mind first principles. We shall refer to 'striking' as the ringing up of the hours (to which may be added a single stroke at the half hour) and to 'chiming' as the ringing up of quarters on a separate train. There are arrangements by which a striking clock sounds at quarters, usually on two or three bells; but such a clock we shall call a 'quarter-striking', rather than a chiming, clock, because it has only one sounding train. Thus a striking clock has a going and a striking train and a chiming clock normally has going, striking and chiming trains. The sequence of events is that in the striking clock the going train lets off the strike; in the chiming clock, the going train lets off the chime and partially releases the strike, but the latter is only fully released by the chiming train at the fourth quarter.

There are two completely different basic striking systems, known as the 'countwheel' and the 'rack', and one or other of them is also used for chiming. In the oldest clocks, the countwheel system was employed, since the rack system was not invented until 1676 and came into general use a few years afterwards. The countwheel was never superseded, however, although it has tended to be used mainly in cheaper clocks in recent times. It involves fewer parts and needs less exact adjustments, but it also has considerable disadvantages. The rack mechanism is essential for clocks where the strike is to be repeated by pulling a cord or pressing a button. In the last two centuries it has been much used to effect the striking of quarters, which can be done more cheaply and simply than by the apparatus of a full quarter-chiming train. On the other hand, it is rather prone to

134

wear out of adjustment over a period, and it is liable to stop the clock if it fails to strike for any reason or is allowed to run down. The two systems are often found in combination, generally in modern clocks, rack striking being used with countwheel chiming. There are many clocks, however, using a rack system throughout, for striking and chiming, and this is usual on the best class of nineteenth-century mantel clock in England, and also on older bracket clocks. It makes no difference whether the clock is weight or spring driven as regards which system is used.

The Countwheel System

The countwheel is a 12-hour (occasionally 24-hour) disc from which notches are cut, the raised sections between them being proportionate to the numbers of strokes required on bell or gong (Fig 44). Thus the basic countwheel is made up of 78 divisions, of which 12 are cut-out slots representing pauses between each hour, for 78 is the total number of strokes struck for all the hours of a 12-hour period. Often, particularly in continental clocks, a countwheel of 90 divisions is used, the slots being twice as large to allow for 12 more single strokes, one at each half-hour. Geared in appropriate ratio (ie 1:78 or 1:90) to the countwheel, starting with a gear wheel which is riveted to it or mounted on the same arbor, there is in the trains of English clocks a wheel on whose arbor is a cam with a deep notch in it, known as the hoop wheel, which takes various forms. In the more modern clocks it is simply a cam with a notch in the edge. In old English clocks it is a strip of metal riveted into the edge of the gear wheel and forming an incomplete circle (the 'hoop'). On some clocks, particularly American ones, a 24-hour countwheel is found and the hoop wheel has two notches in it, but the principle is exactly the same as with the commoner 12-hour system. Pivoted between the plates is a lever having two fixed extensions or detents, one so placed as to be able to drop into the notch on the hoop wheel, and the other so placed as to drop into the notches of the countwheel.

The idea of countwheel striking is that the detent can fall into the notch on the hoop wheel, stopping the train, only if the other detent is able to fall into a slot on the countwheel. Thus, if the striking train is set running at 9 o'clock, the detents are lifted and, as the train revolves, the locking detent goes slightly in and out of the hoop wheel, waiting for a chance to fall right in, the countwheel detent pacing up and down on the raised surface of the countwheel; but the train will not lock again until the one detent falls into the next slot in the countwheel, so that the locking piece is at last able to lodge in the slot on the hoop wheel. Meanwhile, the hammer is raised and dropped nine times, its tail sprung against radial pins of a pin wheel (also called a hammer-wheel). This is next or next but one (according to the duration of the clock) to the barrel or driving pulley, and geared directly with the hoop wheel so that the passing of each pin, or striking of each blow of the hammer, corresponds to the passing of the hoop-wheel notch beneath the locking piece. Sometimes a starwheel is used instead of the radial pins, but the effect is the same.

The details of this system vary considerably according to the date and country of manufacture. In the oldest English clocks the countwheel was mounted on the outside of the backplate (Plate 4). To it was pinned the gear wheel which engaged with a simple pinion on the extended arbor of the driving barrel or pulley, which also drove the hoop-wheel pinion and the rest of the train between the plates (Fig 40). In most 8-day French mantel clocks a similar system was adopted, save that the external countwheel was not driven directly from the barrel outside the plates but was pinned to the square extended arbor of the wheel intermediate between going barrel and pin wheel. In many later clocks, particularly German and American, the countwheel is between the plates. It may be fixed to the wheel arbor or alternatively it may be mounted with a friction washer on a stud in the back of the frontplate – then it itself has gear teeth round its circumference and is turned by an extended pinion trundle in the lantern pinion of the hoop, the latter thus making one revolution for each tooth by which the countwheel is advanced. In these latter clocks the levers are often of steel wire with the ends flattened and are easily bent. Again, in the French pattern (Plate 5) a hoop, as such, was not normally employed. Instead, the corresponding wheel, the locking wheel, has a projecting pin with which the stub-ended locking piece engages (Fig 41) and this method is used in some eighteenth-century English clocks.

136

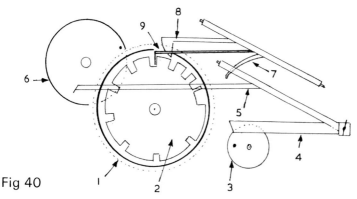

Fig 40

Old English countwheel striking
1. Hoop wheel
2. Countwheel (on different arbor from 1)
3. Minute wheel or cannon pinion with lifting pin
4. Lifting piece
5. Warning piece
6. Warning wheel and warning pin
7. Locking release lever, operated by warning piece
8. Hoop wheel detent or locking piece catching edge of hoop
9. Countwheel detent in notch of countwheel

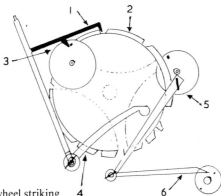

Fig 41

French countwheel striking
1. Countwheel detent going through backplate to lodge on countwheel
2. Countwheel behind backplate
3. Locking piece joined to countwheel detent, engaging with locking wheel between plates
4. Locking release lever operated by warning piece
5. Warning piece, warning wheel and pin
6. Lifting and warning pieces raised by lifting pins on cannon pinion

The striking is set off by a pivoted arm, the lifting piece, which engages, in all save the oldest clocks, with a pin or pins (the lifting pins) projecting back from the cannon pinion or from the minute wheel. Often a small snail piece is used instead of pins for this purpose. The train is released for a moment and then immediately held up just before the hour to ensure that it is free to run precisely on the hour. The pre-run is known as the 'warning', and it occurs because the lifting piece's upper arm not only raises the locking piece from the hoop wheel, but also raises a projecting arm, the 'warning piece', into the path of the second wheel of the striking train, permitting this wheel to run half a revolution (Fig 42). When, on the hour, the lifting piece drops back off the lifting pin, this obstruction is removed and, as the locking was released at warning, the train is free to run. In the old English one-handed clock with external countwheel, a rather different arrangement was made for setting off. Here the hour wheel, driven directly by the going pulley's pinion, was fixed to a 12-pointed starwheel. The end of the lifting piece was an iron curve shaped so that it rode up each tooth of the starwheel, setting off the warning at its

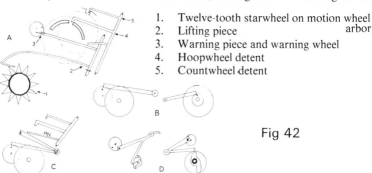

1. Twelve-tooth starwheel on motion wheel arbor
2. Lifting piece
3. Warning piece and warning wheel
4. Hoopwheel detent
5. Countwheel detent

Fig 42

Release of countwheel striking. (For French type see Fig 41 and for cuckoo clocks see Fig 45)

A Old English single-handed clock with striking train behind going train
B Old English using minute wheel or cannon pinion and two hands (otherwise as above). Locking may be by hoop wheel or by locking wheel and pin and locking piece
C Later type, using one-piece warning piece (passing through front plate) and lifting piece. Countwheel may be between plates. Trains are side by side and minute wheel or cannon pinion may be used
D American and general modern types

high point and releasing the train as it fell back into the bottom of the starwheel tooth. It should be said also that old countwheels are not always in the form of slotted discs; sometimes the raised sections are pieces of a broken rim riveted into the side of the gear wheel so as to project, forming an interrupted circle.

The countwheel system, cheap, simple and reliable if properly set up, has one serious disadvantage in that it has a continuous cycle of action. Once the '9 o'clock' section has been passed, next time the lifting piece is raised it will be the 10 o'clock section which presents itself, regardless of what the clock hands may say as to the time. There is then no possibility of a 'repeater' whereby, for example at night, the misheard strike can be repeated at the pull of a cord. This, when seeing the dial was more difficult than merely pressing an electric light switch, was a demerit compared with the rack system. There is also the related difficulty of adjusting the hands, which must be moved round past each hour, stopping to let the clock strike fully before proceeding to the next. Whether through neglect of this fact or for other reasons, such mechanisms do become out of phase, striking a different time from what is shown on the dial, and they will never correct themselves automatically except by making compensatory errors. The remedy is to hold the striking inactive and to turn the hands until they match the strike, or to cause the strike to run until it has caught up with the dial. It is doubtless for this reason that countwheels are so often outside the plates since then, by a hand inserted through the back, the detent can be lifted from the countwheel (thus unlocking the hoop wheel) and the train be allowed to run the necessary amount. Alternatively, a pull string or wire piece is arranged to release the train. I once met an unfortunate who had purchased a good Victorian clock at rather a high price but believed that he had an exceptional bargain. He thought the clock was of earlier vintage, for it was a good reproduction made of nice wood, and the cord, although at present it did not produce exactly the hour required, indicated that the clock was the coveted 'repeater'. Sadly, the cord was the pull-string of a countwheel adjuster.

The simple countwheel system can cause a good deal of trouble, but not so much as the would-be repairer who goes at it determinedly with

Plate 4 English (2-handed) movement with external countwheel

a little knowledge and a file. The great temptation with incorrect striking of this sort is to attack the countwheel, paring off a bit here and punching a bit in there. But the countwheel by its nature is more of a gauge than a working wheel and it is subject to very little wear. If the detent stops the train by jamming against the next section of the countwheel, there is something wrong, because it is the hoop wheel and not the countwheel which should be doing the locking, but the odds are strongly against the fault's being in the countwheel. Often the countwheel's mounting, particularly if it is external, does not correspond in each of its possible positions with the slot or pin in hoop or locking wheel. It is always worthwhile in cases of such difficulty to try the countwheel in different positions on its squared arbor (or in relation to the pinion with which it engages), in one of which all may not be well. Often again, particularly with the flat wire levers, the detent itself is bent – it should not touch either side of the countwheel slots, and its profile at the tip should be radial to the countwheel. The detent should normally fall into the middle of a slot but, where the countwheel has large slots to allow for the striking of a stroke at the half-hour, the detent should naturally fall near the previous raised section after striking the hour, and just in front of the next one before striking the hour. (It should be said that some clocks use the 'hour' countwheel only and arrange for the bell hammer to be triggered separately by a pin on the cannon pinion at the half-hour.)

The interaction of countwheel and hoop or locking wheel is critical. The locking piece and countwheel detent should fall to the bottom of their slots or there is a tendency for the striking to mislock and 'bounce' on to the next hour's striking. The slots of countwheel and hoop wheel must align periodically, as has been said. This is determined by the cutting of the countwheel (and putting it on its arbor or stud correctly) and by the geared connection of pin wheel, hoop wheel and countwheel. The ratio here is one to however many pins there are on the pinwheel (usually thirteen), since the hoop or locking wheel must revolve once for each blow of the hammer, plus whatever further reduction is needed to make the total ratio up to 1:78 or 1:90. With regard to position, the countwheels of older clocks can generally be moved. In more modern clocks, the hoop wheel or countwheel is mounted – and sometimes both are mounted – on their arbors with grubscrews into the collets. Turning and refixing them is then a simple matter. If there is no such facility and the countwheel is between the plates, special care must be taken when setting it up in the first place, or in adjusting later with the plates loosened on the striking side.

There are two important adjustments to be made which are common to all striking systems (Fig 43). One is to ensure that the

141

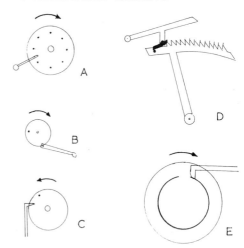

Fig 43

Setting up the striking
A   Position of hammer tail between pins of pin wheel when striking is locked (all systems)
B   Position of warning pin to warning piece when striking is locked (all systems with warning)
C   Position of locking wheel to locking piece as bell is struck (French systems)
D   Position of gathering pallet to lock train (Old English rack system)
E   Position of hoop to locking piece as hammer strikes bell (English and American countwheels)

hammer tail is clear of the pin wheel or starwheel when the striking has stopped. The hammer should have been released and be virtually touching the gong or bell when the train is at rest, otherwise there may be a stroke 'left over' from the previous striking and the wrong number may be struck. It is also desirable that the strike train start into motion without the weight of the hammer and tension of the hammer spring impeding it. If, after striking, you press lightly down on the hammer head and there is any movement of the striking train, the tail must still be on a pin and the position of the pin wheel will have to be adjusted relative to the hoop or locking wheel. It can be a fiddly adjustment, but it is essential to correct striking.

The other essential is to make sure that the train is running freely by the time it is arrested by the warning piece. The warning pin on its wheel should be so placed when the strike is at rest that it has at least half a turn to run before it collides with the intercepting warning piece.

142

It will of course always stop in the same place, since it is geared in an even ratio with the preceding locking or hoop wheel. In many good quality clocks, especially French, the pin wheel and sometimes also the locking wheel are pivoted in subsidiary cocks which can be removed to permit adjustment of these wheels after the movement has been assembled. This is a great help in setting the train. Without it, one can only try to get the positions right first time or persevere with opening the plates and wriggling the wheels round until all is correct, for the striking will be very capricious if it is not. There can be no question of springing the arbors apart between plates and turning the wheels in relation to each other that way.

It not infrequently happens that one has to deal with a 30-hour longcase clock from which the external countwheel and geared connection to the driving pulley are missing. It is quite possible with these large movements to file a pinion and cut a wheel by hand, the size and ratios having been calculated as shown in Chapter 2, pages 39–40. The job can certainly be completed with a home-made countwheel if necessary. For the sake of appearance, this can be cut out of brass up to, say, 2mm thick – there is no other merit in thickness, which will be harder to cut, since the wheel receives no wear. The size can be determined with an experimental cardboard former mounted on the stud (which may itself have to be made) where the wheel will run. The inside radius, from centre to the bottom of the notches, will be set by how far the countwheel detent falls when the train is locked. The outside radius is not very critical, save that the detent, when raised by the hoop wheel, must be well clear of the raised sections of the countwheel. The 78 divisons can be obtained by dividing the circumference of the outside circle (ie radius $\times$ 2 $\times$ 3.142) by 78 and measuring the resulting dimension in steps round a circle. Alternatively, mount a blank disc and mark it appropriately as the detent falls when you strike the clock round. Before cutting the metal disc, the cardboard with divisions marked on it can be stuck to the metal with glue and the centre drilled to fit the stud. Then test-run the model seeing that, as the train is run for each striking sequence, the detent falls on one of the divisions marked, note any modifications and cut the slots (Fig 44). The same procedure can be adopted for a missing French countwheel, though here 90 divisions are required, the notches each contain 2 divisions and are much smaller. They are provided with sloping edges away from the direction of rotation so that there is no possibility of the detent, with its very small movement, sticking on the way out of a notch. In a clock with a hoop wheel, the weight of the detent rests there, but where the locking is by pin (as in most later movements) the weight is usually on the countwheel, which slides beneath the detent. It may be noted also that French countwheels are often interchangeable and a spare one may be available or obtainable from a scrap movement. When a substitute of suitable size is found, it should of course be tried in each possible mounting position and the striking be test-run with it so that any small deviations can be marked on the wheel and adjusted with a file.

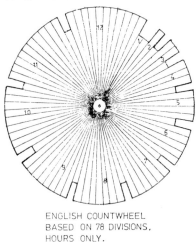

English and French countwheels

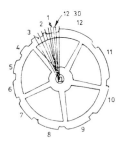

ENGLISH COUNTWHEEL
BASED ON 78 DIVISIONS,
HOURS ONLY.

FRENCH COUNTWHEEL
BASED ON 90 DIVISIONS,
HOURS AND HALF HOURS.

Fig 44

A distinctive arrangement is made in older cuckoo clocks (Plate 6). These may be weight or spring driven and the mechanism does not vary greatly between one or another. They nearly always employ countwheel striking, with the striking train usually, but not always, on the right, seen from the front. Their works are frequently congested and the various levers are wires, so that the apparatus can appear more complicated than it is (Fig 45).

144

Plate 5a French countwheel striking (a) from front (b) from back. Note silk pendulum suspension

Plate 5b

145

Plate 6 Cuckoo clock, bird and striking levers

The countwheel rides on a stud outside the plates (which are usually of wood with the wheels pivoted in driven brass bushes) and is driven by a pinion on the extended arbor of the pin wheel or intermediate wheel. Both locking and warning are usually performed on the second wheel of the train, next to the fly. This has a pin projecting from it and corresponds to the usual warning wheel. The locking piece is on the

Strike and drive mechanism of spring-driven cuckoo clock

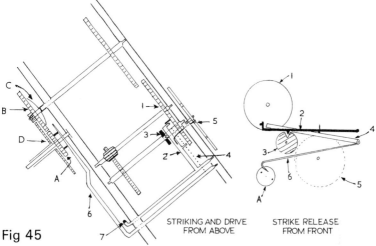

Fig 45

STRIKING AND DRIVE
FROM ABOVE

STRIKE RELEASE
FROM FRONT

1. Locking and warning wheel with pin
2. Locking piece with wire through plate to countwheel, pin to engage with warning piece, and tongue to step in hoop cam
3. Hoop or cam
4. Warning piece, engages with locking piece and pin on locking wheel
5. Countwheel on back of plate engages with pinion on barrel arbor
6. Lifting piece
7. Vertical pin on locking piece arbor operates bird and doors
A Cannon pinion and lifting pins
B Minute wheel and pinion on greatwheel arbor
C Clutch spring
D Hour wheel and pipe

countwheel detent arbor and the tip of it holds this pin when the train is at rest. When the train is released, the pin falls off the raised locking piece but is immediately held up until the hour by a crooked wire, the warning piece, on the lifting-piece arbor. The warning piece raises the locking piece by means of a pin. The locking piece, besides the projection to catch the locking pin, has also a terminal section which locates in a cam (the hoop wheel) when the countwheel detent is in a slot of the countwheel and the train is locked. Sometimes this hoop wheel conventionally locks the train (instead of the pin on the warning wheel mentioned above), but more often it is so shaped as not to hold the locking piece but to cause it, and the countwheel detent on its arbor, to step up and down, waiting until there is coincidence with a notch in the countwheel.

The train is let off by a lifting piece of the curved shape familiar in one-handed 30-hour clocks, working on the usual lifting pins in the cannon wheel. The pin wheel has in effect three hammer tails on it. The first is a true hammer tail in respect of the gong which sounds with the bird. The second and third are attached to the levers which operate, through wire links, the low and high noted bellows for the bird's song. The hammer usually sounds first, but this is a matter easily adjusted by bending the hammer tails to change their effective length and position. The high note must, of course, sound before the low note, so it is essential to connect the levers to the bellows the right way round. A final lever, bending along the top of the movement, is pivoted into the wall with a spring, so that it is normally turned slightly backwards. To this lever is attached the bird, which is so contrived that raising its tail will open its wings, this being done by a short wire which rises and falls as the low-note bellows open and shut. The end of the lever, or the feet of the bird, have a wire hook or hooks linked to them and these are arranged so that when the bird moves forward it opens the little door or doors in front of it. This movement is effected by a vertical extension of the locking-piece arbor; when the train is unlocked, the locking piece rises, the extension moves sideways and flicks a pin in the lever on which the bird is mounted, pushing it sideways and forcing the bird forwards (Fig 46). Needless to say, the levers must be so adjusted that the doors do not open or the bird come forward at the warning, but only when the train is fully freed. This is simply a matter of clearance between the pin on the locking-piece arbor and the cranked lever to which the bird is attached.

Perhaps because the bushes are mounted in (often poor quality) wood, or because the tendency has been to try to produce these clocks as cheaply as possible at any one time, or because they struggle on for years under adverse conditions – for whatever reason, these clocks seem to present themselves in an advanced state of wear, particularly in the pinions and pivots. The wear is usually general and it is scarcely worthwhile remaking such clocks; they are in the most literal sense

Hammer and bird mechanism of cuckoo clock

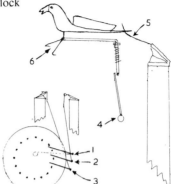

Fig 46

1. Gong hammer
2. Low note hammer
3. High note hammer
4. Vertical pin on locking piece arbor operates bird and doors
5. Wire from low note bellows operates bird's tail and wings
6. Hooks on bird support connected to doors

worn out. Still, there are basic adjustments which can be made. An important one is to the locking of the striking train, ie the setting up, as already noted (Fig 43). With three 'hammers' it is all the more important to ensure that all are 'down' when the clock has finished striking, and to make sure that they do not rise during the run to 'warning'. Another factor is the fly or fan. This is mounted with a flat spring to give it some play when the train suddenly stops running. Too loose a fit here, and also in the pivots throughout the striking train, can cause the striking to rush, with an increase in the chance of mis-locking. Then attention has to be given to the levers to ensure that they do not cause unnecessary strain on the workings and that they move as far as intended. A point to watch here is the length of the wire connections to the bellows; if these are too long, it will be impossible for the pin wheel to release the tails of the 'hammers' and the striking will stop midway. Finally, the bellows and pipes must be replaced or remade if necessary. New pipes are often available from suppliers, but pipes can be made up to the pattern of the old from strips of veneer or from the wood of a cigar box. They should, of course, be of thin wood, and glued, not nailed, together. Bellows can be replaced with a section of kid glove or of an airtight synthetic material, after carefully removing the old remains and laying them out as a pattern. The

149

bellows, again, are stuck on. It is never satisfactory to try to patch or stick old bellows, which usually start to leak at the hinge and then go at the folds. Birds can be bought — they can also be usefully touched up with paint. Sets of plastic figures in various sizes for the applied dials can be bought and have only to be stuck on. The modern material is whiter than on most of these clocks and, for appearance's sake, it is usually a case of replacing a complete set of figures rather than a single digit.

The Rack System

Rack striking was in general use by the late seventeenth century in England. It rapidly replaced the countwheel system as the standard system of striking in the quality clock, though countwheels continued to be more widely used in other countries. The clear merits of the rack were that the same strike could be repeated by releasing the train and that, provided it was set up correctly, the hands and striking could never be out of phase. The reason for this is that the number of strokes struck is dependent on the position of a stepped cam (the 'snail') which is either solid with the hour wheel or is mounted free but turned directly by the cannon wheel. The complete strike for, say 8, is as it were contained in the eighth step on the snail and this will always be in the same position at 8 o'clock.

The striking train is linked to the snail by a curved rack with ratchet teeth. This rack is pivoted on a stud so that its angled lower arm (the rack tail) can fall onto the snail by gravity or by means of a spring. The distance which the rack can fall is governed by the position of the snail. One of the wheels, the locking wheel, has an arbor extended to the front and on the squared end of this is fitted a single tooth, in modern clocks a brass disc with a raised eccentric pin, known as the gathering pallet. Its purpose is, by revolving, literally to 'gather' the rack up tooth by tooth until it can go no further. One stroke is struck on bell or gong for each rack tooth gathered, for the locking wheel here is geared to the pin wheel in the same way as is the old hoop wheel in the countwheel system. The rack is kept from falling back as it is gathered up by a pivoted lever, an unsprung click, which catches in the rack's teeth and is known as the rack hook.

The old English and the continental (and modern) systems are rather different in their arrangements for locking the train and holding the rack up once it has been gathered (Fig 47). In the English system the gathering pallet has an extended tail opposite the pallet. Until the rack is all gathered, this tail can rotate with the pallet, but once the rack is gathered the tail collides with a projecting pin at the head of the rack and the pallet can no longer turn. The pallet is then horizontal, free of the rack, which is held up by the rack hook. In the more common arrangement the rack hook is curled in shape with an angled end, though there are various forms. Whilst the rack is being gathered past it, the hook is of course displaced to the left, but when the rack is fully gathered the rack hook drops in beneath it and holds it raised. This rack hook is mounted on the same arbor as the locking piece, which moves in and out with the hook according to whether it is displaced or in stopped position. When the hook falls into place the locking piece also moves in and its blunt finger intercepts a pin on the locking wheel so that the train is halted. In the modern clock there may be no locking wheel, so that the locking piece acts on the warning pin in the warning wheel. Alternatively, locking may be by the shape of the brass gathering pallet, which has an angled edge to coincide with a projection half-way up the rack hook and which is locked when the two meet – only when the rack hook is in the stopped position.

The snail of older clocks is often mounted not on the hour wheel but on a starwheel. This starwheel has 12 teeth and is advanced one tooth for each revolution of the cannon wheel, by means of a pin on that wheel. (Thus the snail can in fact be turned independently of the going train when it is not actually engaged by the pin but, as the starwheel is provided with a jumper and spring, this cannot happen by accident.) The starwheel's jumper ensures that the wheel flicks swiftly into position when it is advanced. The purpose of the arrangement is to delay the change in the snail's position until immediately before the hour is struck, rather than to have it slowly changed by continuous gearing as happens if the snail is mounted directly on the hour wheel. The quick change is particularly desirable for repeating clocks which otherwise would be liable to strike the next hour when their repeating work was operated, before it was strictly due (Fig 48).

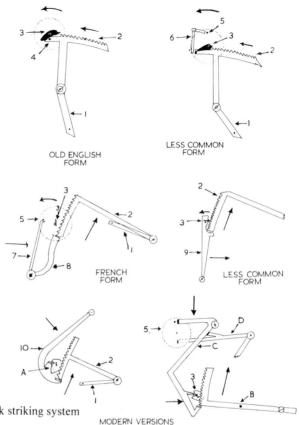

Fig 47

Locking of rack striking system

MODERN VERSIONS

1. Rack tail
2. Rack
3. Gathering pallet
4. Pin on rack obstructs pallet tail
5. Locking wheel and pin between plates
6. Lever moved by rack pin, and locking piece between plates
7. Locking piece between plates
8. Rack hook on locking piece arbor
9. One-piece rack hook and locking piece whose pin obstructs pallet tail, all on front plate
10. One-piece rack hook and locking piece obstructs pallet cam
A   Pallet cam with pin as tooth
B   One-piece rack and rack tail
C   One-piece rack hook and locking piece, whose tail goes between plates to obstruct locking wheel
D   Warning piece works on same wheel

152

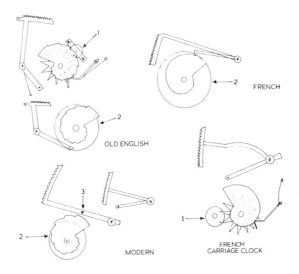

Fig 48

Racks, star wheels and snails
1. Cannon pinion with pin turns star wheel
2. Snail fixed to hour wheel
3. Pin on rack acts as rack tail

The rack has normally more teeth than are strictly required for hourly striking and its first tooth is often deliberately short. This short tooth is used for the half-hour striking (not usual on classic English clocks but common on continental and more modern ones). There are two lifting pins on the cannon pinion (or on the minute wheel), one to operate the hours and one to operate half-hours. The latter is nearer the arbor and consequently causes the lifting piece and rack hook to rise less far and only partially to release the rack, so that the first, short, tooth is engaged by the gathering pallet, but no more. This arrangement does not wear very well although, within limits, the short tooth can be stretched out to compensate for wear. In due course it may be necessary to dispense with half-hour striking on such a clock, and then the additional teeth at the top end of the rack are useful since 12 can be struck without the need for a new rack. It is also of course possible to dovetail in a new half-hour tooth, the process being similar to that described (Chapter 2, page 42) for fitting replacement teeth into a gear wheel or, often, suitably to relocate the half-hour lifting pin.

153

A different and preferable arrangement for half-hour striking involves the release of the train whilst the rack is stopped from falling by an ancillary lever operated by the minute wheel. Then the gathering pallet merely picks up the bottom of the rack and sets it down again, whilst one blow is struck.

The setting off device for rack striking (Fig 49) may be much as that in the countwheel system, although it again differs in English and continental models. In the traditional English arrangement (Plate 7)

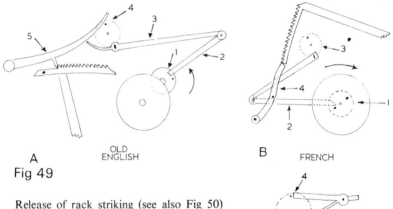

A   OLD ENGLISH   B   FRENCH

Fig 49

Release of rack striking (see also Fig 50)
(*above*)
1. Minute wheel and lifting pin
2. Lifting piece
3. Warning piece, obstructs warning wheel and releases rack by raising rackhook
4. Warning pin
5. Rack hook, by releasing rack, unlocks train which is held by pallet tail on rack pin
(*centre*)
1. Cannon pinion
2. Lifting piece engages with pin in rackhook to release rack
3. Warning pin and warning piece
4. Rack hook, on arbor of locking piece between plates, frees train
(*below*)
1. Lifting cam below cannon pinion
2. Warning piece with extensions which (3) obstruct warning/locking wheel and raise rack hook
3. Acting edge of warning piece
4. Locking piece extension of rack hook releases train when raised
5. Rack hook

the minute wheel has the same number of teeth as the cannon pinion and has a lifting pin, or two lifting pins diametrically opposite. As this wheel revolves, its pin raises the lower arm of the lifting piece, and the upper arm raises the rack hook. The upper arm of the lifting piece is extended back into the movement and this extension, the warning piece, catches the warning wheel's pin as the train starts to run. Meanwhile the rack tail, helped by a spring, falls onto the section of the snail opposite it. When, on the hour, the lifting piece drops back off the lifting pin, the warning is released and the train runs, the gathering pallet starting to raise the rack tooth by tooth.

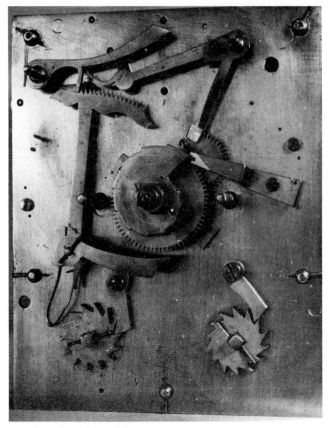

Plate 7 Old English rack striking

In the continental pattern these parts are in different places since the rack falls vertically and the rack hook drops in below it (Plate 8).

Plate 8 French rack striking with lifting-piece release

The lifting piece is raised by pins, usually mounted directly on the cannon pinion, and it lifts a pin in the rack hook, which releases the rack. Rack and rack tail are at a lesser angle to each other than in the old English form, and they are pivoted on a stud on the right-hand side of the plate. The rack hook and lifting piece are placed on the left. Subject to this rearrangement, the action of warning and release is as in the English version. Modern forms are nearer to the continental, but the rack hook, which varies in shape, is often suspended from above and has, as mentioned, a bent projection to engage with the cam of the gathering pallet (Plate 9). Locking may be by this means or by a similar, but reversed, projection back between the plates to operate on the warning-wheel pin. On striking clocks, the rack is usually to the right, where there is a space, but on chiming clocks it is placed near to the traditional 'English' position, or near the centre of the plate, since the right-hand space is occupied by the chiming mechanism. Placement does, however, vary considerably.

Plate 9 A modern English rack-striking movement

For rack striking there exists a completely different means of setting off. It is very common on carriage clocks, especially repeaters (Plate 10). This is the flirt mechanism, 'flirt' being an old word for a lever giving a sudden jerk, and curiously connected with the same word used of transient amatory affairs. The flirt (Fig 50) is a jointed arm pivoted at the top. One part of it is shaped down to a point and is pushed back against a spring by the lifting pins as they go round. The other part is mounted loosely on the main arm about half-way up and held with a downward tendency by a weak spring mounted also on the arm. This cross-piece has at the end a deep notch or hook which, when the flirt is pushed back before striking, drops down onto a pin behind the extended rack hook and lodges there. When the lifting pin has gone far enough, the flirt is released and flies across the clock from front to left. In so doing, it causes the notch on the loose arm to push against the pin on the top of the rack hook and to move the rack hook aside so that the rack can fall. The moving of the rack hook of course also releases the locking piece and, as there is no 'warning' in this arrangement, the train begins to run. The gathering pallet has a second and larger tooth behind the usual one, and this is used to knock the flirt's notch off the rack-hook pin once the rack begins to rise. In this system the snail is always mounted on the sprung starwheel so that it snaps round just before the hour. Sometimes the flirt has two notches, one so placed that it moves the rack hook enough to release the train but not the rack. Thus 'one' is struck for the half-hour as usual. This arrangement depends on careful positioning of the pin at the top of the rack, and is often unreliable. Locking piece and rack hook may also need adjustment. The flirt system, with the absence of warning and the use of the starwheel, was ideal for, if it was not actually invented for, repeating clocks, and it is general, though far from universal, on French repeating carriage clocks.

In the setting up of the rack strike, as of the countwheel system, the same two principles are of prime importance (Fig 43). It is essential that, if there is a 'warning', the warning wheel has room to run, and the hammer must be off the pin or starwheel when the clock has finished striking. The latter condition will be met, and is most easily adjusted, if it is ensured that, when the hammer has just struck, the

158

Plate 10 French rack striking with flirt release

locking wheel has about one-eighth of a turn to run before it meets the locking piece, or before the English pallet tail hits the stop-pin in the rack or before (in a common older form) the French pallet tail hits the stop pin in the rack hook. The gathering pallet should be so placed that it has at least a quarter of a turn to run before engaging with the rack, and thus is usually placed at 9 or 10 o'clock on its arbor; this is particularly important with the flirt system where there is no warning and the rack must have fallen fully before the pallet starts to gather it up. The pin wheels of many old French clocks are extremely small, and indeed the whole mechanism is fine and closely adjusted, and it is vital that they be properly set up in this way — though fortunately most rack striking clocks of this type have the sub-cock for the pin wheel, already mentioned. Another guide is the maker's punch-mark between two teeth on a wheel — I say 'the maker's', because a subsequent repairer's scratch or dot may not be reliable and will be detectable by its shallowness. Between two teeth thus marked will go the pinion leaf whose corner is ground off at an angle to distinguish it.

159

Fig 50

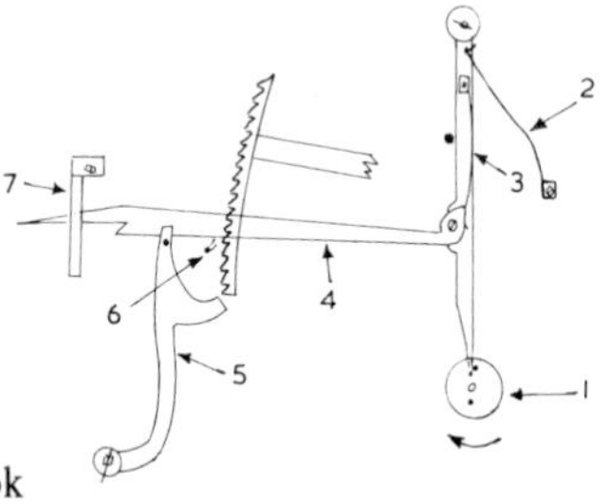

Flirt striking release (shown after striking)
1. Cannon pinion and lifting pins
2. Spring tensing flirt against stop to left
3. Spring tensing flirt down onto rack hook
4. Flirt ending in notch which engages with rack hook pin when lifting pin forces flirt to the right
5. Rack hook on same arbor as locking piece between plates
6. Gathering pallet whose larger tooth (nearest plate) disengages flirt from rack hook pin when train runs
7. Forked guide for flirt

The rack system, despite its clear advantages, is more easily upset by wear and abuse than is the countwheel system. One of the chief difficulties can be wear (or, more often, wrongly suspected wear) on the rack itself, which seems to cause the striking to be unreliable. Obviously the gathering pallet must collect the rack teeth with absolute reliability, or the sequence will go entirely wrong. If it catches on the corner of a tooth or repeatedly drops one so that the rack is never picked up, something must be wrong. In the case of the repeatedly dropped tooth, provided that it is always the same one, it is usually found that this tooth is worn and can be slightly stretched with a punch. Where you have a less regular mistake the difficulty is almost certain to be not in the rack teeth but in the rack tail. On old English and many other clocks this is connected frictionally with the rack proper so that the two are adjustable at the centre. This adjustment can be used to govern which tooth is presented to the rack (for example, if the shallow half-hour tooth is removed) or to alter the engagement of the pallet by slightly raising or lowering the rack in relation to it. If these parts are fixed they can, more drastically, also be bent or filed with similar effects, but absolute certainty as to the cause of the trouble is needed before one starts to take metal off a rack, for the adjustment is less easy to reverse.

Where there is a flexible rack tail, or a rack tail with a pin on an inserted spring, its operation should be checked. It is a protective device

160

so that, if the clock fails to strike, for example because of a broken line or through not being wound (perhaps deliberately by the owner), when the snail comes round to 1 o'clock it is able to push aside the rack tail rather than breaking it or stopping the clock. (Striking clocks which stop between 12 and 1 o'clock are always suspect here.) Check that the spring is not too stiff and that the engaging edges of the snail and rack tail are chamfered and polished. It may be noted in passing that many modern rack striking clocks with the snail on the central arbor are so disposed that they use a one-piece rack and rack tail – the rack is of inverted shape, the toothed section projecting upward from the arm, and as a result a sprung pin can be inserted midway down the arm to act as the rack tail. (Examples are shown in Fig 47 and Plate 9.)

In adjusting the rack tail it must be ensured that the rack will fall sufficiently far for a full twelve teeth to be gathered, and it should be checked that the mounting itself is properly firm, for if the angle can alter during the running of the clock the position of the striking will clearly be untrustworthy. A similar point to be checked is the operation of the lifting piece and warning. The two arms of the lifting piece are not normally friction-set, but they can of course be slightly bent, and it is necessary to make sure that the warning pin does not slip on the warning piece when that is raised, and that it fully clears the warning piece when that is down. So far as the older rack hook is concerned, the blunt-nosed profile is essential to the proper holding of the rack teeth as they rise, particularly so with vertical racks. Though it may be worn, it will not normally be worn all over, so that its original shape can still be discerned. With the badly worn hook one may either stretch it and file it back to its true shape, or add hard solder and file it back to the shape of the unworn section. Mere filing must be avoided, for a short rack hook will upset the locking of the train, except in the old English system, and cause the rack to slip.

Where the hourly sections are marked on the snail in steps, the rack tail should strike midway, in the middle of each step. When they are not marked, extra care must be taken to see that 12 o'clock, (half-past 12) and 1 o'clock are correctly struck. If they are not, the snail may have to be turned back slightly and it will be necessary to check all the strikes through to determine its best position.

161

In the old English form the 'strike-silent' mechanism can give trouble. It consists essentially of a flat spring over the end of an extended back pivot of the lifting piece. By virtue of this, the lifting piece is normally held forward so that it does not engage with the lifting pin on the front of the minute wheel, and the striking is silenced. By means of a lever with an angled surface, the lifting piece can be pushed back against the spring so that it is raised by the lifting pin and the strike is set off. The lever is connected to an indicating hand, usually in the break arch. What has to be ensured is that, once the lifting piece is pushed in, to 'strike', it stays there and fully engages with the rack hook and lifting pin, and also that the spring really does push the lifting piece sufficiently far forward to 'silence' it when the indicator hand with its cam is turned. As has been said, it is not really satisfactory to silence a rack striking clock by merely neglecting to wind its mainspring. The only adequate arrangement is to leave the train wound and to raise the lifting piece out of the path of the lifting pins, and this is the normal procedure with the various strike-silent levers that are employed.

As one receives clocks less countwheels, so one may receive (but mercifully less often as these are more internal pieces) clocks less rack or snail. It is no easy job to make up one of these (let alone both), but it can be undertaken in a worthwhile cause. The simplest procedure is, as with making a countwheel, to run the train through its full sequence, this time marking on a cardboard disc (for the snail) the points to which the rack tail falls, to give each number of strokes. For example, when the pallet is in the second tooth, mark a circle on the card corresponding to where the rack tail falls, and so on with the other teeth. The uprights of the 'steps' can then be drawn in, one in each circle on twelve radial lines, and the snail provisionally cut, leaving the edge somewhat proud for adjustment. It is possible to reverse this process for a rough 'dummy' rack, marking where the teeth of the rack fall as its tail touches each division of the snail (or where each division would be, if there are none, as the hands are turned). In practice the teeth will be apart by a distance just minutely greater than the size of the pallet tooth itself (ie, excluding its arbor). It is usual for there to be thirteen or fourteen teeth. The length of the rack tail is the same as the distance between rack pivot and centre arbor, since its action should be radial. The length of the rack itself can be gauged from the space between its stud and the pitch circle of the gathering pallet. The curve of the rack will be that of a circle whose radius is the distance from the rack's stud to the space between the seventh and eighth teeth of the rack. Thus the rack can be roughed out on card as a triangle, the three points being the rack stud and head and foot of the rack, and the curve will be a circle whose centre is the rack's stud. The depth of the teeth will be appropriate to the gathering pallet and they may now be cut out and the pattern tried and adjusted before it is fixed to metal and sawn to shape.

Developments of Rack Striking

The most far-reaching extension of the rack striking principle is into various forms of 'quarter striking', where the quarters are struck by the hour train, not chimed through a separate train of wheels. (Such an extension was also made to the countwheel system, but it is extremely rare.) There are two distinct methods, one being the doubling of the rack and the other being the complication of the snail.

So far as the complication of the snail is concerned, a disc is used from which at appropriate intervals three downward steps are cut, the fourth and final level being that of the conventional snail (Fig 51). Thus at 12 o'clock we find first three steps, representing the quarters of the hour, and then a very deep slot representing the most cut-away part of the hour snail. There are four lifting pins stepped out at varying degrees on the cannon wheel, and the first three teeth of the rack are graduated short. The fourth lifting pin raises the lifting piece completely and releases the rack for the hour strike. There are usually two bell hammers, though there can be more. That for the higher pitch is on a sprung arbor and is pushed back away from the pin wheel at the hour by a lever operated by a pin on the minute wheel. Thus usually the deeper note alone sounds at the hour. A rather different arrangement is occasionally met in which a pivoted lever rests on a quarter snail driven by the minute wheel, and this lever has three steps on its end. The bottom of the rack misses this lever at the hour, but at quarters the quarter snail places the lever in the rack's way and its

163

movement is then limited to one, two or three teeth. This is a more reliable mechanism than that of the compound snail, for the latter can present difficulties with the rack tail not falling reliably onto the correct quarter step.

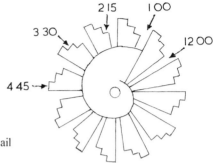

Fig 51

Combination hour and quarter snail
for quarter striking 'ting-tang'

It is possible to use graduated lifting pins together with graduated teeth at the lower end of the rack. Hence the commonest form of quarter striking (outside carriage clocks, which use a different system) merely involves two bell hammers, one retracted by a lever at the hour, and four lifting pins, placed so as to progressively release the rack, except at the hour when it is fully released and the tail falls onto the snail. There is a potential problem around 1 o'clock, when the snail's edge will be so high that it will prevent almost all movement of the rack, whereas two and three teeth have to fall for the half-hour and three-quarters striking. This is solved by making the high '1 o'clock' section of an ordinary hour snail shorter than usual, and following it by a notch sufficient to let the rack make its modified fall for the quarters until 2 o'clock, when the 'normal' snail section reappears (Fig 52).

The essential points in adjusting all these quarter-striking arrangements (over and above the general rules for setting up striking) are to examine with care the placing of the lifting pins and the reliability of the engagement with the rack's lower teeth, and to make sure that the lever which disables the second hammer on the hour does in fact move up at the correct time to do so – this is a matter of adjusting the motion work's position and is often assisted by punch-marks of the manufacturer on the motion wheels. The lever should be

164

in position to silence the high-noted hammer when the hour strike's warning is given. The whole striking work is most easily assembled in position just before striking the hour and then a check is made round the striking to see that all is well.

Simple 'ting-tang' quarter-striking (shown striking half-past one)

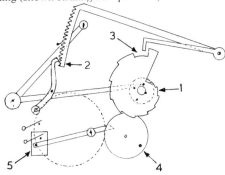

Fig 52

1. Cannon pinion with graduated lifting pins
2. Three bottom teeth used for quarters and hours
3. Lowered section of snail to enable quarters to be struck after one o'clock
4. Pin on minute wheel operates hammer silencing lever at the hour
5. End of hammer silencing lever which raises one hammer tail from the path of the pin wheel at the hour

Quarter-striking carriage clocks work on a different principle from the above in that they use two racks, one for quarters and one for hours, but the quarter rack is behind the hour rack and the teeth of these racks coincide. There is only one rack hook but it has a higher and a lower foot, the latter for the quarter rack, and so shaped that it cannot hold the quarter rack unless the hour rack is fully gathered up. There is the usual hour snail mounted on a starwheel with jumper, and also a quarter snail which is mounted on the cannon pinion – the pin on this turns the hour snail's starwheel just before each hour (Fig 53).

There is no lever from the minute wheel to prevent one of the hammers from sounding at the hour. (In some complex and valuable carriage clocks there are as many as four hammers and bells or gongs.) Instead, there is a contrivance more easily illustrated than described (Fig 54). A pivoted arm in its raised position allows both hammers to sound, but when lowered, as it is on the hour (since its

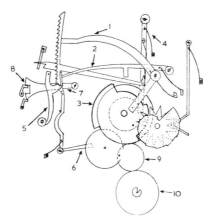

Fig 53

Main front-plate work of quarter-striking carriage clock
1. Hour rack. Tail falls on snail which is advanced by action of pin of cannon pinion on star wheel (see Fig. 48)
2. Quarter rack. Tail falls on quarter snail attached to hour wheel. At hours, both racks fall and coincide, one hammer being silenced. At quarters the hour rack is prevented from falling
3. Quarter snail
4. Flirt (see Fig. 50)
5. Rack hook
6. Pivoted lever preventing hour rack from falling. Operated by minute wheel
7. Pin on hour rack holds piece raised to permit both hammers to sound unless hour rack is released (see Fig. 54)
8. Pivoted piece controlling hammer action
9. Intermediate wheel. Set-hands wheel below (not shown)
10. Alarm wheel (see Fig. 38)

support is a pin at the foot of the hour rack which will have fallen), it obstructs one of the hammers, leaving the other free to strike on the deeper bell or gong. To prevent the reverse situation (ie of the hour sounding at the quarters) a lever is held by a spring so that it will not permit the hour rack to fall. For striking the hour, this lever is moved aside (like the 'silencing' lever in the quarter-striking clocks just described) by a pin on the minute wheel, and then the hour rack is allowed to fall. The quarter rack cannot fall (though released by the rack hook), because it is obstructed by the highest section of the quarter snail. One hammer is silenced, as above. There is a gathering pallet of double depth to pick up the teeth of both racks – in fact, it is of triple depth, for there is also the usual large pallet tooth at the back which separates the flirt from the rack hook.

With this arrangement, quarters can be struck on two or more bells and hours on one bell. If the clock has a suitable mainspring and appropriate gearing, it will also be fitted with a hand lever to adjust to silence, 'petite sonnerie' or 'grande sonnerie'. 'Petite sonnerie' is the normal quarter-striking action of quarters on two bells, without hours save at the hour. 'Grande sonnerie' is the sounding of the previous hour at each quarter, thus telling the full time to the nearest quarter of an hour. To obtain grande sonnerie, the hand lever sets aside the sprung lever which normally prevents the hour rack from falling at the quarters. When the hour has sounded, its rack pin raises the small pivoted lever again so that, after a silent 'dummy blow', both bells can sound, the quarter rack is gathered and the quarters are struck.

It is of course most important with these clocks that the cannon pinion (and quarter rack) and hour snail are correctly positioned, for the clock should, if a repeater (as these clocks invariably are) strike the preceding hour and three quarters until only a minute before the next hour is due, and the hour snail therefore must not be able to turn early. For silencing the striking, the hand lever usually has a third position in which it prevents the flirt from falling onto the rack hook, so that the latter is not displaced and the train is not released.

The qualification of 'a suitable mainspring and appropriate gearing' for these clocks is important. The simple quarter-repeating carriage clock cannot be modified merely by inserting a hand lever so that the hour rack can be brought into action at any quarter. The trains of a true grande sonnerie clock and a quarter repeater are quite different in proportion, the grande sonnerie needing higher counts and a larger barrel and mainspring. A faked clock is most unlikely to strike satisfactorily on a grande sonnerie basis throughout the week, even if stopwork on the striking barrel is discarded. Fakes are not uncommon, for the grande sonnerie has attained a value and a demand which make the label worth acquiring – although one may harbour the personal feeling that the difference in price between the two types of clock tends to be out of proportion to the normal

Hammer control in quarter-striking carriage clocks

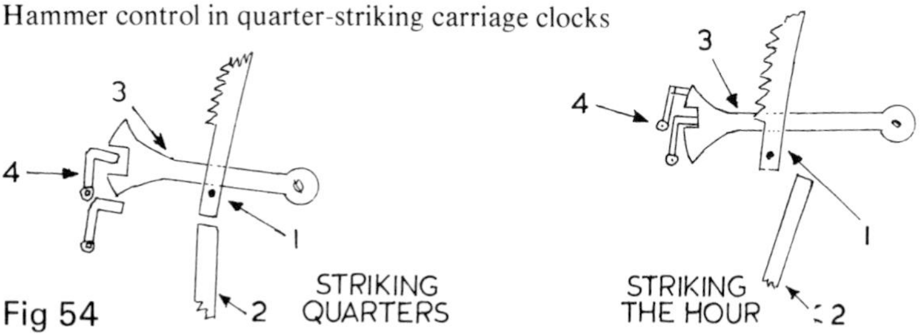

Fig 54   STRIKING   STRIKING
          QUARTERS   THE HOUR

1. Pin on hour rack holds slotted lever up when rack is raised during quarter-striking
2. Lever operated by pin on minute wheel only permits hour rack to fall at the hour
3. Lever controlling hammers. The high-note hammer is obstructed when the hour rack and the lever fall for striking the hour
4. High-note hammer

differences between their movements, and I have yet to discover the owner of a grande sonnerie clock who was prepared to listen to the full striking performance day in and day out.

The repeating mechanism of a carriage clock consists of a short lever connected to the press button and having two forks (Plate 10). One of these goes through the frontplate and terminates in a pin level with the fly, ensuring that the striking will not start until the button is released. The other fork descends to the rack hook, which it pushes aside. In quarter-striking clocks, it normally also pushes aside the hour-rack retention lever so that, at least at repeating, the hour as well as the quarters will be struck. Earlier repeating systems, not used in carriage clocks, are straightforward enough to repair as a rule and are seldom met with, other than for repairs to the simple lever and cord for releasing the lifting piece. In some now sought-after pieces there was a small chiming train controlled by a quarter snail revolved hourly by the minute wheel. These were wound to each occasion by a pull-string and unlocked the hour train so that the hour could also be sounded. Such clocks are unlikely to come the way of most amateur repairers, and any which do might in humility be placed elsewhere.

The Comtoise or Morbier clock (named after its place of origin in the Jura) presents several curious features, including a unique rack striking system. These weight-driven clocks were made for over 150 years until early in the twentieth century. They are attractively odd, even the relatively modern ones, and so they are in demand with collectors, who require them set up and adjusted. The movements are very well made, as a rule, with heavy well-cut wheels. The clock's form differs little as between old and modern, save that the verge escapement, mounted upside down, was used in the older clocks, which struck on bells, whereas a good recoil escapement and striking on gongs became general later. There is a quaint contrast in these clocks between their elaborate decoration and their large-scale open movements, which have something of the turret clock about them, and operate with a similar clatter. The pendulums, like the surround to the dial, are massive in proportion but light in weight, for the multifarious, apparently carved, work is in fact embossed on thin brass (and frequently has been put straight and soldered from behind) and the enormous brass discs of the pendulum bobs are hollow, not lead-filled. The cases vary greatly, though the pendulums and weights are usually open to sight and air. Often there is no case. Movements were evidently sold uncased, even at a late date, and then cases were made if required. Consequently, it is the name of the case-maker or retailer which usually appears on the dial and neither the name nor the style of case may have anything to do with the Jura region.

Comtoise movements (Plate 11) are posted, like the old English lantern and 30-hour clocks, and encased in removable iron panels. Strip plates are used at back and front for each train, and again in the middle for the motion work. The pendulums are suspended centrally just behind the dial, in front of the movement, and feature a circle in their rods to bypass the motion work (Fig 55A). The going train and escapement are on the left and the striking train, as is usual in the posted clock, is on the right when seen from the front.

Once one is familiar with the appearance of these clocks one will be prepared for the oddity of their striking arrangement (Fig 55B). A straight vertical rack is used, toothed on both sides and placed at the front of the movement, where it slides down by gravity when released. The rack hook operates on one side and the gathering pallet on the other. These clocks strike once on the hour and once again two minutes later, the strike being exactly repeated. This double strike is

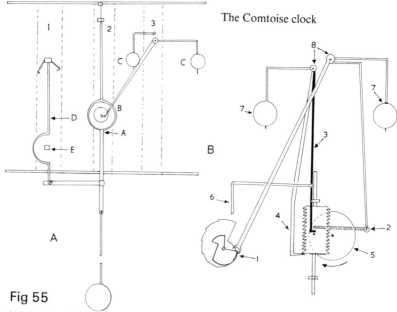

The Comtoise clock

Fig 55

(A–Outline of movement.
1. Going train strip plate
2. Motion-work strip plate
3. Striking strip plate
A Pendulum rod with clearance for centre arbor
B Lifting piece, hour wheel, snails
C Lifting piece counterpoise and rack hook/locking counterpoise
D Lateral crutch
E Going barrel square
B–Striking mechanism, shown after first striking at hour
1. Lifting cam and double-ended lifting piece behind hour snail
2. Pivoted locking release lever operated by lifting piece, at rear of movement
3. Locking piece, at rear of movement
4. Rack hook
5. Gathering pallet at front, locking wheel at rear of movement
6. Wire rack tail
7. Counterpoises
8. All parts on same arbor are fixed and move together

due to the lifting assembly, which is held engaged not by a spring but by two prominent counterpoise weights as it rides on an arbor high above the train to the right. At the dial, the lifting piece has a forked end and presses against a cam of unusual shape, hidden behind hour

Plate 11 A typical Comtoise movement

wheel and snail. When the arbor has revolved to an hour, the first prong of the lifting piece falls off the cam so far as the second prong will permit and the clock strikes. Two minutes later, the cam having revolved a little further, the second prong also falls off, so that the clock strikes again. If this action is to be reliable, it will be seen that the distance between the two prongs must be half the radius of the cam from its centre to its highest point, and it is most unwise to attempt to adjust the lifting fork unless it is plainly incorrect from this

Plate 12 A modern countwheel chiming movement

point of view. At the half-hour, the cam is shaped so as to allow only one prong to fall off, so there is no repetition of striking, and the usual arrangement is followed whereby the rack is not released for the single half-hour stroke. Between the hours, the rack is held up by a long rack hook, and the train held by a locking piece on a locking pin; rack hook and locking piece being again on an arbor high over the train wheels and with a counterpoise instead of a spring to ensure their engagement. The train is released by a pivoted lever on the lifting-piece arbor. This lever moves across the locking wheel at the back of the movement and displaces the locking piece from the locking pin. The rack is gathered in the usual way by a pallet on the arbor of the locking wheel. The rack's fall is limited by the hour snail (which is screwed to the hour wheel and is usually of hollow, cut-out form), and this is contacted by means of a wire arm bent round the strip plate and fastened on the top of the rack, both rack and arm falling vertically. This wire arm corresponds to the normal rack tail. Finally, it should be noted that there is no 'warning' in this action.

The two enemies of these fascinating clocks are excessive (especially thick) oil and worn pivots. The rack is not sprung, but falls by its own weight (which is small) when released. Its lateral movement is limited at top and bottom by a pin moving in a holed brass bracket. As there is not the usual 'warning' between release of the rack and the full running of the train, it is essential that rack and arm (rack tail) fall swiftly onto the snail before the gathering pallet has had time to begin collecting the teeth of the rack. To ensure this, the rack's bearing pins should not be oiled, but checked for roundness and smoothness, and the holes be bushed if necessary. It is the course of desperation, but possible if the rack can still be lifted easily, to use a little lead to add weight to the rack. It has also to be ensured that the gathering pallet is properly placed when the rack is fully gathered – it must have a good run to make before it starts to gather, to give the rack time to fall.

The pivots and linkages of the lifting piece and also of the rack hook and locking piece must be adjusted until they are free, but with little shake, so that their counterpoises are positively effective. If necessary, the gathering pallet has to be replaced – it will not be satisfactory merely filed true but short, and the engagement cannot be easily altered by moving the rack sideways. It goes without saying that similar attention must be given to the lateral crutches of these clocks – sometimes they jam and sometimes they have so much shake that the pendulum loses impulse through the rather devious transmission. A new suspension spring is often needed, for these pendulums can be difficult to free from their hook just below the dial, with the result that the spring, concealed behind the dial, becomes distorted, causing such a pendulum as this to yaw badly and, very likely, to stop. The other common site of serious wear is, as with 30-hour English clocks, in the hammer arbor, which is held in line with the hammer pins against a strong spring. The striking will suffer if this spring is bent to slacken it. Proper bushing of the hammer-arbor holes is the best, and the craftsman's answer.

Chiming

The classic old English chiming system illustrates clearly the basic arrangement by which the going train lets off the quarters and the fourth-quarter chime lets off the hour. The chiming train is the same in principle as the rack-striking hour train (Fig 56). Four lifting pins on the minute wheel raise a lifting piece which in turn lifts a rack hook and releases a rack, in this case usually having five teeth, one for each quarter and one for the rack hook to rest in. The lifting piece is attached as usual to a warning piece stopping a pin on a warning wheel between plates. The rack tail falls onto a quarter snail which is attached to the minute wheel and, according to how far it is permitted to fall, a pin wheel (or a pin barrel similar to that in musical boxes) revolves and releases hammers onto gongs or bells (or indeed, a choice of either). The usual pin at the head of the rack is engaged, when the rack is fully gathered, by the extended tail of the gathering pallet, so locking the train.

So far all is as with the old hour rack striking, save for the different position of the train which obliges the rack hook to be lifted by a pin on the warning piece rather than by its tail. The difference lies in the connection of this train with the striking action, which must occur only after the fourth quarter and must both release the hour rack and set the hour train at warning so that it will run at the right moment. This is accomplished by having the quarter rack sprung so that it moves (unlike the hour rack) from right to left when released – moves, that is, towards the striking train. The hour rack is released only by the fall of the quarter rack at the fourth quarter, which is deep enough to permit its head to knock a pivoted lever, the other end of which is the hour rack hook and which, when so jolted, releases the hour rack. The connection to the hour warning wheel is by another pivoted lever which is lifted by a pin (the stop pin for the pallet, passing right

174

through the rack) at the head of the quarter rack when the quarter rack is fully gathered. Thus the hour warning piece comes into position whenever the quarter rack is released, but the hour striking cannot run until its rack hook is released at the fourth quarter, when the warning wheel will engage with the warning piece until the quarter rack is fully gathered up, at which point the warning piece falls and the striking runs. Only at the fourth quarter is there the combined release of the rack followed by the release of the warning which will permit the hours to strike after the chime has sounded.

Similar points of adjustment apply as to the hour train. The chime hammers must be free of their pins when the chiming train is at a halt, and the quarter warning pin must have room to run before contacting the chime warning piece. The pin on the rack and the gathering pallet must be adjusted so that the train locks reliably and the angle of the rack tail must be such that four quarters on the snail are definitively engaged. What, however, must also receive attention is the linkage with the striking train, for there must be no chance that the quarter rack will contact the striking train's pivoted rack hook before the fourth quarter and, when the fourth quarter arrives, the quarter rack must infallibly release the striking train and rack. The hour warning piece has to be adjusted so that the train is released when the quarter rack is fully gathered, but its spring must not be so stiff that there is doubt in the gathering of the quarter rack. The biggest factor in these adjustments, which are after all mostly to very old movements, is the fit of pivots in their holes. It is useless to expect consistency of action where there is great shake in pivots or where the threads of screw studs for levers and racks are stripped.

There is an alternative method of release in these movements, using a flirt. Here there is no 'warning' in the chiming train. A sprung lever is flicked by the lifting pins and its cranked end rises and knocks the rack hook upwards so that the quarter rack will fall and release the train by freeing the gathering pallet from the pin on the rack. This is a considerable simplification and presents no special problems since with the English locking system by the pallet tail, the pallet must, if it locks, always have sufficient run to allow the rack to fall when there is no 'warning'. Trouble can arise in the flirt's spring. This may

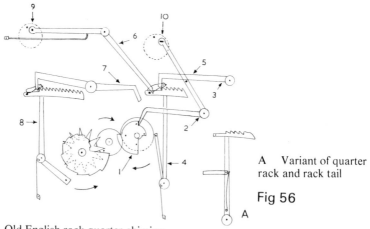

A   Variant of quarter rack and rack tail

Fig 56

Old English rack quarter chiming
1. Quarter rack with lifting pins on minute wheel
2. Quarter lifting piece and warning piece
3. Quarter rack hook raised by pin on warning piece (or by lever on warning piece stud)
4. Quarter rack
5. Quarter warning piece and pin to lift rack hook
6. Hour warning piece, sprung, operated by pin on quarter rack
7. Hour release lever and rack hook operated by fall of quarter rack
8. Hour rack
9. Hour warning wheel and pin
10. Quarter warning wheel and pin

have to be bent so that it is strong enough to raise the rack hook, but not so strong that the encounter of flirt and lifting pin stops the clock.

The movements employing rack striking and chiming on this basis are generally good examples of what may be called the classic English tradition. With three fusees, large spring barrels, heavy plates and wheels in proportion, they have a solidity which is impressive and, to the addict, incomparable. The fact remains, however, that they are necessarily very heavy and rather cumbersome. They suit the monumental single clock of the household rather than the clock which happens to rest on the mantel-piece of humbler folk. Unfortunately, reduced in size and with thinner metal, the system loses its characteristic merits and develops the wear and widespread maladjustment which is such a menace in mass-produced rack movements with an economical use of metal. These faults are

176

particularly prevalent in the much used quarter train. In point of fact the full rack chiming clock was gradually replaced by the movement with rack striking on the hours but a countwheel system for the quarters. This is the norm now for chiming clocks, and has been for nearly a century. (The application of the countwheel to quarters was not of course a new invention. There exist early clocks with large countwheels subdivided so as to take account of the quarters.)

The operation of such a typical modern chiming clock is still clearly related to principles already noted. In particular, the completion of quarter chiming is employed to set off the hour train, and quarter chiming by countwheel can become out of phase unless some corrective mechanism is built in. (Hence earlier clocks with count-wheel chiming are provided with the customary cord or wire for letting off the chime until the phasing is restored.) The lifting pins are normally replaced by a starwheel mounted on the centre arbor, and it is common but not universal for one of the teeth of the starwheel to be extra-long for letting off the hour train and self-correction provision. The two commonest lifting arrangements differ somewhat from each other (Fig 57). In one (Plate 12) there is a long lifting lever engaging with the starwheel's teeth and raising the countwheel detent, and it has on its arbor the warning piece engaging with the warning wheel. A projection of the countwheel detent backwards into the movement constitutes the locking piece, which acts on a hoop wheel. Thus, when the lifting piece is raised, the quarter train is unlocked by the countwheel detent's locking piece, but is immediately held at warning. When the lifting piece drops off the starwheel the train is finally free to run and will do so until the detent drops into the countwheel slot, and the train locks as the locking piece coincides with the notch in the hoop wheel. In the other arrangement a short, right-angled lifting piece is employed, being pivoted on a stud above the central arbor and with a light spring to keep its lower arm in the path of the starwheel – being also reversible in movement so that the action will not be damaged if the hands are wrongly pushed backwards. When it is turned, this lifting piece raises an intermediate lever, an extended warning piece, and also raises the countwheel detent which, having a locking piece on its arbor, unlocks the train. As before, once the train is at warning, the countwheel has advanced so far that the countwheel detent cannot fall back into a slot, nor its locking piece engage, so that when the warning is released by the falling of the lifting piece, the chiming has to continue until countwheel and locking or hoop wheel again coincide. Often the warning piece has a pin pointing outwards from it and it is this which raises the countwheel detent.

In both types, of which there are of course various versions, the

Modern forms of countwheel chiming and strike release

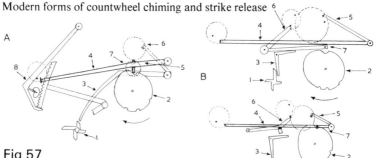

Fig 57

A   Strike flirt has projecting locking piece (working on hoop wheel) and countwheel detent. The locking piece goes back to the hoop wheel between plates

B   Locking is by detent on strike flirt arbor. The chime warning is operated by the lifting piece from the left and it raises the strike flirt and countwheel detent

1.   Lifting cam on centre arbor
2.   Countwheel with raised fourth quarter to release hour rack
3.   Lifting piece
4.   Strike flirt, which may have projections or pins acting as locking piece and countwheel detent
5.   Locking piece, which may act on locking pin or on a hoop wheel
6.   Warning piece and chime warning wheel
7.   Countwheel detent, which may merely be a pin on the strike flirt
8.   Hour rack, released by high rise of strike flirt on chiming for fourth quarter

countwheel detent corresponds to the quarter rack-connecting lever in rack chiming, in that it is extended right across the clock so that the left-hand end can release the hour train by letting off the warning and, progressing a little further, can release the hour rack, the train being held at warning. The hour train cannot then be fully released until the countwheel detent falls into a notch on the countwheel and the clock has finished chiming. This long detent, often known as the strike flirt, can only release the hour train at the start of the fourth quarter. This is ensured by having the fourth-quarter section of the countwheel of a high, bumped shape (or, sometimes, by having a raised ridge in the normal section). It is only when the strike flirt is raised extra-high by this protrusion that it can touch the hour rack hook.

The varieties of self-correcting device are legion and, because with a little study their action will be clear, there is no point in enumerating them at length. What has to be remembered is that the striking is actuated at the beginning of the fourth quarter by an abnormality, mentioned above. The principle of self-correction is to lock the quarter train after the third quarter more deeply or specially, so that only the higher lifting action required to release it can unlock the train at this stage. As a result, if through some jolt or (more commonly) advancing of the hands, the clock strikes the third quarter when it should have struck (for example) the first, the chiming train will be held arrested after the third quarter and the wheels will not be released until the higher action is produced as the minute hand approaches the hour.

Common arrangements for self-correction are shown in Fig 58. One set-up employs a supplementary locking plate or countwheel with one notch, screwed by bush and grubscrew (so that its position can be set) onto the countwheel arbor and either alongside the countwheel or between the plates. On this additional cam rides a further detent which will lock the train if its end is in the notch, but is free of the wheels if its end is on the rim of the cam. The notch is aligned with the third-quarter slot on the main countwheel and when this detent locks it can only be released by the action of the longer starwheel tooth in raising the strike flirt higher at the fourth quarter. Thus, as the starwheel is fixed to the central arbor, only the fourth quarter and the hour can be struck when this longer lifting tooth approaches '12' on the dial and the chiming, if out of sequence, will be arrested after the third quarter until the minute hand reaches the hour. The additional locking is carried out by a pin on the locking wheel or by a second 'hoop', again usually adjustable by grubscrew, on the same arbor. A second, very common, device is a spring clip mounted on the back of the countwheel in the slot before the fourth quarter (Fig 58B). This clip is

179

caught by a projection or pin on the lifting lever as the countwheel revolves for the third quarter, and it holds the countwheel immovable until the strike flirt is raised the extra distance needed to set off the hour train at the fourth quarter, when the catch springs back out of the way. If this extra lift is not forthcoming at the next raising of the lifting piece, the chime train cannot move until the hands 'catch up'.

Regarding countwheel striking movements, we have noted that it is essential that the countwheel be set up correctly, or the locking and the slots in the countwheel may coincide only irregularly and the striking will be most peculiar. The same applies to the chiming countwheel. On modern clocks it is usually mounted by bush and grubscrew so that its position is readily adjustable. The hoop wheel or notched cam of the locking is very often, but not always, similarly mounted; thus it is not too difficult to arrange matters so that the locking piece is properly engaged in the locking slot when the countwheel detent is properly in a countwheel slot. (This adjustment should be made regardless of the noises emitted by the chiming at this stage, since the chiming barrel is separately adjustable to match the stopping points decided for the train.) Associated with the placing of the countwheel is the placing of the self-correcting locking cam, if there is one in the clock concerned. This must be such that the auxiliary locking piece drops firmly into the notch when the countwheel is locked before the fourth quarter. The auxiliary locking is usually set very slightly in advance of the main locking, but not so far that the main locking is at all hindered. The other point requiring special attention is, as in the full rack chiming system described above, the linkage with the striking train at the hour. The strike flirt must be so shaped or bent that it releases the hour warning and, at the hour, the hour rack, but not the hour train until the detent is back in the countwheel slot. The warning pin has as usual to be positioned to permit a free run, but this cannot be as extended as in the striking system, and between an eighth and a quarter of a revolution is the normal setting. If too long a run is given, the chime is likely to emit one or two noises during the warning, depending on the gearing to the chime barrel and the closeness of the pins on it.

However well the chiming train is set up, it will be of no avail if the chime for a particular quarter produces the 'wrong' tune or, worse, two parts of a tune. This is, so long as the hammers are free when the train is locked, entirely a matter of the position of the chiming barrel. To establish this position can be very difficult in old clocks chiming on bells, for the tunes are not of recognisable shape to the modern ear, but are in fact based on the changes of church bell-ringing. Often, too, the difficulty of obtaining the correct position, even if known, is considerable, for the wheels gearing the main chiming train to the barrel may all be between the same plates on an old clock and it is necessary at one and the same time to make the adjustments for full rack chiming as well as for correct position of the barrel. In modern clocks, the difficulty is less, since, by and large, standard tunes, with well-marked cadences, are employed, and the ratio wheels and barrel are usually mounted separately outside the backplate, and moreover fitted with an adjustable collar to the turning arbor of the train.

There are ten sequences of chiming in an hour, since the chimes are cumulative (the third quarter, for instance, consisting of three bursts of chiming). It would not be practicable to use a barrel on which all these chimes were set out in sequence once only. In practice, therefore, chime sequences are always divisible by two, there being a complete cycle for the first five sequences (that is, up to and including the second sequence of the third quarter) which is repeated for the second five (beginning with the third sequence of the third quarter). The chiming barrel thus has to revolve twice for each revolution of the wheel in the chiming train, with which it is geared, and on whose arbor the countwheel is placed. The locking-wheel pinion has a count one-tenth of this wheel to make its ratio 1:10, since each revolution of the locking wheel accounts for one sequence of the barrel. It follows that it is possible to set the barrel up with the 'first' tune (often easily recognisable as a descending scale and straight line of pins, though of course this depends entirely on the tune used) as the first quarter of the hour or as the last sequence of the third quarter of the hour. Naturally, in practice the one to choose is the first quarter of the hour, for then the train can be at rest immediately before this more identifiable part of the barrel is positioned.

If, as often happens, a chiming clock comes in where the chiming

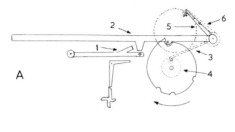

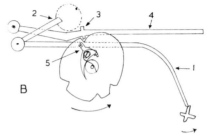

Fig 58

Two common self-correcting countwheel chime mechanisms. (A–Subsidiary countwheel and locking piece B–Countwheel catch (shown from behind)) (*above*)

A
1. Chime warning piece, also raises countwheel detent pin out of main countwheel, lifting it higher, by a longer lifting cam, during the fourth quarter, so releasing hour rack
2. Strike flirt with strike warning piece and countwheel detent pin
3. Subsidiary countwheel detent, only locks and releases at fourth quarter. This detent is free on the arbor of the strike flirt
4. Subsidiary countwheel with one notch placed at fourth quarter
5. Main locking piece fixed on strike flirt arbor
6. Subsidiary locking piece, solid with (3), free on strike flirt arbor. In the first three quarters, this piece is held away from the locking pin by the action of (4). At the fourth quarter, when it has fallen, it is raised by the main locking piece, which acts on a pin

(*below*)

B
1. Lifting piece with pin
2. Warning piece on arbor of lifting piece
3. Locking piece, a projection on the strike flirt, operating on a hoop locking wheel (not shown) between plates
4. Strike flirt with locking piece and countwheel detent projections
5. Sprung catch at rear of countwheel. The hook engages with the lifting piece pin so that the catch is pushed in line with the fourth quarter countwheel notch. The train is then locked until the long lifting cam finger raises the lifting piece high enough for the pin to clear the catch, which can only happen at the start of fourth quarter chiming

182

barrel is obviously in the wrong place, it pays first to identify on the barrel the intervals between sequences (where there will be fairly observable lines across the barrel without pins in them), and then to arrange the barrel temporarily so that it at least produces correct sequences which may run on and be stopped at the wrong places. Finally, listen to these sequences and from what you hear pick out what appears to be the initial sequence, and mark on the barrel where it begins. The barrel may then be repositioned, and the train run till it reaches the first quarter, so that this sequence may be tried out as the first. The chiming is then run over the hour, trying out other 'first' positions if necessary, until the correct, or at least the most acceptable, arrangement is found. Several manuals set out in musical notation the commoner chime tunes. For the very old ones which can hardly be identified and often have ill-defined pauses, the best that can be done is to settle for the most pleasing effect.

Chiming barrels vary in construction. The oldest, best, and most expensive, is akin to that found on old musical boxes – a tube of hardened brass into which are driven steel pins. This is still used in good-quality modern clocks, particularly where a choice of chimes or tunes is required. (Such choice is given by a pumping action, which may be actuated by the clock itself, moving different circles of pins into line with the hammers, and it does of course require extreme accuracy in the spacing and action of the hammers. The beginnings of the sequences correspond, but there is little room for adjustment and the barrels must be very carefully placed. Sometimes separate barrels and hammers are used for different tunes.) These barrels can be repaired by driving broken pins through into the inside and replacing them with pins as similar as possible, taking especial care that the protruding length is correct. They are best held with the arbors spanning a notched open vice or two blocks of wood, because it is easy to damage more pins than are repaired.

Suited as it is to the finer action, the traditional cylinder is expensive to produce, particularly in quantity, and there have been many substitutes, especially where the chiming requires only four hammers (as does the Westminster chime). The commoner varieties are a steel drum with pushed-up snag teeth – strong but virtually impossible to repair in the event of breakage or serious wear – and a composite barrel made of a row of steel starwheels with spacers between them. These wheels are of course stamped out and are cheap to manufacture. They are normally driven hard onto the arbor so as to be virtually solid, with brass spacers. Sometimes there are locating lugs on wheels and spacers, which prevent disaster if the assembly comes adrift – if there are not, the best hope is to compare a barrel of any type which plays the same tune. It is always wise to draw or scribe a line straight across these barrels before dismembering them so that there is less difficulty in reassembly. It may be worth noting that the same problem occurs with some elaborate quarter-striking and grande sonnerie carriage clocks. Here, where there are four or more hammers, there is the same number of fine pin wheels mounted with spacers between the plates, and vitally dependent on their relative positions for the correct tune. If you are presented with a broken one, do what you can, of course, but otherwise leave the pin-wheel assembly alone, and do not be tempted to try to improve the regularity of striking by adjusting the position of the wheels – the clearances are fine and the distance between a reasonable strike and chaos is uncomfortably short.

The musical clock is not easily distinguished as a type from the chiming clock, which can, after all, play long melodies. The term is most appropriate when the 'music' dominates the time-indicating function and there is at best a tenuous proportional relationship between the tune and the period (for example the tunes at a quarter past the hour and at three quarters past the hour) to which it refers. The music is neither that of bell changes, nor that of tunes (Whittington, Michael, etc) by tradition and structure suited to chimes which tell the time, but rather a tune such as 'Greensleeves' or a hymn tune, thought to be of perennial appeal as an occasional background noise. Hence the term is used especially, but not solely, of those clocks in which a lever releases a separately wound movement playing on a 'musical box' metal comb or carillon the same tune every hour until a different tune is chosen by the owner, and whose 'music' could be operated manually at any time to delight the ears of those who knew not cassette recorders and transistor radios.

Frequently, whether or not of this type, and whether or not combined with a more conventional quarter-chime system, the tune is played by the clock at intervals of little use for regulating one's affairs, for example every three or four hours, and there is simple stopwork to prevent it from being played more than once. Sometimes a different tune is played for every day of the week and a 'sacred' melody for the Sabbath. The music may be wound from the front and the barrel project from the rear, or be parallel with the backplate (in which case it can be long and accommodate many combs or hammers), or the whole may be situated in a separate compartment in the base of the clock, as is normal with skeleton clocks and was the most usual practice with French clocks. These arrangements do not present any special problem if the principles of quarter chiming are understood.

The repair of the large combed 'musical box' movement may well be better entrusted to a specialist when the clock and its chiming have been repaired and adjusted to set it off correctly, for it is very easy to make matters worse with an elaborate barrel mechanism, and making good for an accident will not be cheap. There is, however, an important thing to be learned from the hammerless action and that is the critical importance of the comb teeth and the pins, not only to the tuning but also to the action. The pin-barrel which operates a comb is analogous to the pin-barrel which operates most quarter-chiming. The friction between the pin and the hammer, as between the pin and the comb-tooth, is a material factor, and so is the great leverage involved where the hammer is concerned. The hammers, on their common arbor, are engaged by the pins very close to that arbor, whilst the hammer head is at the other end of a long wire. It requires a large force to displace the hammer for but a very short distance, as a result of which the hammer head rises up for, perhaps, a couple of inches. This calls for fine adjustment of the nibs of the hammers with which the pins engage. After a few years they become trenched with wear and therefore the hammer is not lifted as it was intended to be, though the barrel of course has an easier journey. This wear has to be taken out and the end of the hammer very slightly bent to restore it to its best distance from the pins. When the chiming is locked the hammer nibs, disengaged from the pins, should be in a dead straight line, or

185

otherwise they will clearly not all receive the same lift. The bells or gongs also should be aligned, particularly if they are tuned rods. This is especially important if, as in the common arrangement this century, it is devised that at the hour an arm connects some or all of the hammers so that a chord is struck.

The scope for repairs to sounding devices is rather limited. A broken tongue can be replaced by soldering a new one on and filing it to tune, but it is hard to produce a tone of the right character. A broken or cracked bell must be regarded as hopeless; although it can be soldered for appearance and to produce some sort of noise, it will never ring satisfactorily and there is really no choice but replacement. The same applies to the occasional bell which has not been drilled dead centrally. Bells which − often with minor adjustment to the mounting − can be fitted are usually available, but to replace, for example, the very flat bells used on the backplates of some bracket clocks can be difficult, since movements of these clocks have long since ceased to be available as a source of spares and are in great demand. Modification may, unhappily, have to be made to the mounting, moving the bracket to one side so that a deeper bell can be fitted without its contacting the pendulum or crutch. The small bells of French clocks, on the other hand, are readily available, and movements can still be obtained reasonably cheaply if an otherwise unobtainable bell is needed. Nests of bells (most often eight) for chiming are more often missing than incomplete. They are hardly to be found. One is not likely to be presented with a clock minus all bells for repair, and if anyone likes to buy a clock for himself in such a condition he should know what he is taking on. Careful attention should be paid to the condition of the spacers between these bells, for often the sound of the nest will benefit from their renewal.

The coiled gong, which made its appearance in about 1870, is more often bent than broken. Unhappily, one condition is almost as serious as the other. A gong which is bent beyond the point where it will spring back is liable to produce only a dull thud, however hard the leather on the hammer head, and there is really no remedy but to replace it − though some owners are horribly casual about the sound which they will tolerate from their clocks. New gongs are available and there is also a good chance of finding a suitable one secondhand. The coiled wire is soldered to a block which may either be the top of the mounting standard or be screwed tightly to the standard or the backplate. It is risky to attempt to re-solder a gong, as the temper at this mounting-point is essential to the tone. A gong which is loose here is better wedged with a pin driven in and sawn off where it emerges from the block, or stuck with epoxy resin. A rusty gong can be cleaned with fine emery, though of course it must not be uncoiled in the process. Many of these gongs are blued to a deep colour and this may be important where they are visible, as in many carriage and skeleton clocks. They cannot be reblued with heat after cleaning, for this would risk the temper, and painting with a blue steel enamel, whilst it sometimes works, is liable to spoil the tone and to look wrong for a high-quality clock. Where a good blue is essential in a rusted gong, replacement is the best answer.

The tuned rods of most modern underslung or longcase chiming clocks can be bought individually or mounted in their standard which, being a large pipe, has a good deal to do with the quality of their sound. It is also possible to buy untuned rods and to tune by cutting, using a piano, pitch pipe, or other tuning standard. (Buy larger than original length.) At the point of mounting, the rods are thin and brittle. If a rod is bent, but sounding, it is better to adjust the hammer to it than to take the risk of bending the rod. The ends are tapered screws jammed tightly into the standard. They cannot be satisfactorily fitted in any other way and a broken rod entails replacement by a new one complete with screw. Rods can be lowered in pitch by cautiously filing them just outside the point of mounting. More drastically, they can be raised in pitch by shortening. Normally they enter the pipe through a hole which does not quite touch them, cross the inside, and are screwed into the other side of the pipe. A rod which touches the outer hole must be bent, despite the risk of breakage, because it will cause a most unpleasant vibration, as will pieces of dirt trapped between rod and hole. Naturally, all mountings of gongs and bells must be firm and free from vibration. Very often an improvement in tone can be obtained by judicious insertion of large washers between the mounting-point and the case.

On all the sounding devices the condition and action of the hammers is obviously important and the craftsman will strive for the best possible sound, regardless of the indifference of his client. Metal-headed hammers should strike bells as near to the rims as possible. Sometimes the hammer falls on a bent wire or metal spring, but as often the spring in the rod of the hammer itself is relied upon to prevent jarring. Either way, the head must stop short of the bell and spring against it and sharply away again so that there is a clean strike. The spring (which takes many forms) impelling the hammer may be adjusted so that the hammer moves with maximum force consistent with not impeding the train. These springs are not a legitimate means of adjusting the speed of striking. This can sometimes be done by modifying the power supply and the fly, but not by means of the depth-setting screw often fitted to the fly pivots, for maladjustment here will only lead to trouble with the pinion in the long run. Radical change can only be brought about by altering the gear ratio of the warning wheel and fly pinion, an alteration not authentic in an antique clock and not likely to be worth the trouble in a modern one. Gongs are struck by hammer heads lined with leather to prevent clanging. The state of this leather has a considerable effect on the tone of the striking and what should be done, if anything, depends as much on taste as on the individual clock. Leathers can be softened by deep pricking with a pin and hardened by singeing and gentle hammering. Suitable leather can of course be bought for replacements, but one soon acquires a stock of old hammer heads from discarded movements, and these will do as well. Hammers should strike near to the gongs' mounting-points and they must strike in the middle of the metal. It is often necessary to introduce shim washers onto the arbor of an old set of chiming hammers to restrict their sideways shake, which is greatly magnified at the end of the head.

7 EXTERNALS

The Case

Although no longer so important to the household, nor so dominant in a room, the clock still remains in a modest way a piece of furniture and, no matter what its mechanical merits, if it is not acceptable to individual taste it will not be tolerated. In the past, craftsmen of immense skill and also designers, as evidence proves, devoted much thought to matters of proportion and decoration, striving to give the clock an aesthetic appearance which neither impaired its efficiency as a machine, nor detracted from its function of telling the time on a dial. Those who made clocks attempted, as clock-makers do today, to bring out the best in the materials available to them, and the repairer must try to do the same. Considering the multiplicity of materials which have been used, books could be written on the subject of clock cases alone. But the case has also considerable bearing on what goes on inside it, and it is partly from this point of view that we have to think of it.

For the longcase clock the essential is firmness. The combined weight of pendulum, driving weight and the very heavy movements commonly built is such that the long case itself cannot give that firmness, except perhaps in the instance of some of the very massive cases used in the nineteenth century. The long case can be viewed as what in origin it probably was – less of a support than a cover for pendulum and weights, the movement being hung on the wall. These cases should be screwed to the wall or, if this is undesirable, they should at least be pressed hard into the wall by thin wedges driven under the base at the front. The backboard ought, of course, to be vertical, so where there is a skirting-board to be considered a batten of the same thickness should be fixed behind the backboard to make sure that it is firmly against the wall without leaning back. Many backboards have seen better days – the necessary screwing does in time help them to split, and they are often badly warped. Ideally, they should be replaced with well-seasoned wood of the same thickness. Such timber is not always easy to come by, however, and the backboard, though invisible, is so large a part of the original case that people are loath to see it disappear with all its marks of antiquity. One may then have to compromise and fix two or three stout 2in battens across its width, if possible outside so that these battens may serve the purpose of balancing the skirting board at the same time, but if necessary inside, in such a way that they do not interfere with the hood. With regard to the skirting, it seems an act of vandalism to cut the base to fit, but it should be noted that there are well-authenticated cases where the base is a front with tapered sides, partly perhaps to meet this problem and partly to ease the sweeping chore of an earlier age.

The earliest longcase hoods slid up – the cases were little more than 6ft high. The countwheel striking would require adjustment from rear or side, the winding was by pulling the weights, and setting to time with the single hour-hand was probably only done periodically, so there was but limited requirement for access to the movement from the front. This form of hood did, however, continue in many instances with key-wound clocks and rack striking and with minute hands, though gradually the hinged door and hood sliding forwards became general even for clocks with the older style of mechanism, which were made throughout the eighteenth century. The forward-moving hood was locked by a sprung bolt, accessible only from inside the (locked) trunk door and engaging with a slot in the hood. The necessity for all this security has aroused speculation, but it seems best to assume that it was because of the clock's great value as timekeeper for the household, which we find hard to appreciate today; if an early clock were tampered with, correcting it was less easy, and the household relied on it to govern their affairs. The upward-sliding hood had a sprung wooden catch (the spoon) which was pivoted inside the top of the trunk so that the lower end sprang forward when the trunk door was opened, and the upper end jumped backward, releasing the hood.

These gadgets, like bolt and shutter maintaining power, are not often found intact, and it is a matter of taste whether to attempt to restore them, after consulting a good model in a museum.

Repairing a longcase (in the sense of squaring up its angles, replacing or battening split wood, regluing some joints and pressing down occasional lifted veneer) is part of clock-repairing. It merges, however, with restoration where, for example, a turned pilaster must be replaced in style from the right wood, a base panel needs to be re-veneered harmoniously, or the question arises as to whether the original colour should be brought out (so far as it can be judged) or defective marquetry replaced on a large scale. There is, of course, an awkward boundary line where replacement with a modern brass hinge may be unnoticed and the substitution of a modern finial may seem justified. But in general I doubt whether the clock-lover who has no very specific experience in cabinet-making and other required skills should venture into restoration, though if he chooses to run the risk of spoiling his own clock that can only be his affair.

We can, in this do-it-yourself age, be a trifle immodest and a little over-catholic in our enterprise. The fact is that even in the golden age of English clock-making, say from the mid-seventeenth to the mid-eighteenth century, horology was a specialist activity with firms and families making parts of clocks rather than the whole machine, and if the assembler and coordinator whose name appeared on the dial was fully qualified in every branch of the activity, that is not to say that he generally made complete clocks and cases in every detail from start to finish. There is a point when the amateur repairer, particularly if he has a notable clock and a paying client to deal with, should admit that someone else would do a better job than he, if his customer requires the best. If he has done good work so far, he will be respected for that advice.

This is still more true of the older bracket-clock cases. There were of course country cousins, but the fullest arts of the cabinet-maker in the choice and finish of woods and decoration were exercised on these clocks, many of which, from their elaboration and contemporary documents, are known to have been precious investments from the start. Their smaller panels make the problem of restoration appear less great than with longcase clocks, but in fact any minute blemish seems that much larger. Nevertheless, bracket-clock cases, because of their value and because they were simply less of an obstacle around the house, have generally suffered from less severe buffetings than longcases. It is quite within the average repairer's scope to replace lost fillets around possibly missing glass panels, to replace the panels, to file up an old key to fit a lock when its pattern has been discovered by dismantling it, to replace the silks behind the brass or wooden frets, and so on. These are useful repairs which cannot violate the appearance of the clock (though the owner should be consulted before gaudy silks are applied) and which must help in the major objective of excluding dust.

When it comes to finials, cast handles, caryatids, lockplates and the like, the matter will depend on what is available at any particular time, on the taste of the repairer and on that of the owner. The choice is often a difficult one between replacing with a modern set because one item is missing, or leaving the set incomplete because of its original character. (This arises especially with dial spandrels.) It may, of course, be possible to have a missing piece specially cast, using the existing pieces as models. Failing this, a mould can be made from thin sheet-metal beaten closely round the object, into which molten scrap-lead is poured. This casts easily and is simply worked on afterwards if detail needs highlighting. The leaden substitute can then be painted with suitable gold-leaf substitute, or true gold leaf and size may be applied. It is also possible to mould in plasticine or plaster of Paris and to cast in one of the modern resin fillers frequently used for car-body repairs and modelling. Of course these methods must not be used for possibly weight-bearing items like handles. Ornament is usually half-shaped with a flat back for mounting. If necessary, however, rounded pieces can be cast using a separate mould for front and back and then fusing the two.

It frequently happens that a clock case has had one side roasted in the sun for years whilst the other side has hibernated and preserved an apparently original rich dark colour. As with restoration generally, the total refinishing of a valuable case is not really within the repairer's province unless he has special knowledge and experience. But, provided that drastic sanding and stripping are avoided, he should surely do what he can to make the clock as a whole presentable, at least so far as the owner will agree (for many owners believe a little verdigris suggests age and value). An effective overall cleaner is a mixture of linseed oil, vinegar and pure turpentine. The vinegar will remove dirt and the linseed oil will tend to darken the wood. This mixture can be used harmlessly over the whole case, and repeated applications will restore the faded area, the vinegar being reduced and later omitted. This is particularly suitable to the marvellous, slightly dull gloss of old waxed wood where the colour and depth are produced by years of rubbing, usually with a wax polish, and will be utterly ruined if sanded down and stripped. Many so-called antique wax polishes can be bought in the shops, or you may prefer to make your own, experiementing with melted beeswax and real turpentine – the proportions vary, but the easier the polish is to apply the less effective it is likely to be and a polish which has to be rubbed on really hard (so long as it is not so solid as to leave lumps in crevices and mouldings) is best. Such polish is applied, at least to complex surfaces, with a stiff brush to even it out, being afterwards rubbed with a wool-free rag. Ebony or ebonising on the older cases can be touched up with matt blackboard paint and then polished overall to match. The use of black gloss paint is vandalism.

The many more modern finishes, varying from true French polish to cheap varnishes on plywood and plain oiled hardwoods, naturally require different methods. Where the skating-rink finish of French polish is involved, the work of refinishing on a valuable clock is usually for the specialist. On a less valuable clock, where the polish has been dotted with water or scratched, some repairs can be undertaken. The surface may be cleaned and improved by rubbing on, with a linen fad, a mixture of real turpentine, vinegar and methylated spirits in equal parts. A motion of overlapping circles is used, as with actual French polishing, to avoid streaking along the grain. It is impossible to touch in a French polished surface satisfactorily with a patch or line of new polish unless the whole area is afterwards treated with a reviver of this sort to even it out. Scratches can be touched in with button polish, which is akin to French polish, comes in several shades, and may be applied with a brush before going over the whole. Where the finish is not high, rings, spots and damp-marks can be levelled back by taking down the whole area with steelwool and linseed oil, following up with reviver to which has been added a very little neutral-coloured French polish. All these operations must be carried out in dry and warm surroundings or the polish will end up streaky, and care must be taken that the rubber or fad is recharged immediately it starts to become tacky during the work. As a general principle, to which the superficial scratch is an exception, it is far better to alter colour by staining raw wood or using a coloured filler than to attempt to make good the colour with layer upon layer of tinted polish. The modern oiled finishes of natural wood present no problems, save that the repairer will have to keep the clock for some weeks if he is to make up for years of neglect. When scratches have been filled, linseed oil is rubbed in hard almost daily and, providing that the underlying surface is good, a dull sheen will eventually result. The use of teak oil is not confined to teak and it has the advantage that it dries more quickly and fewer coats are required, but it will not darken a light wood to the same extent if that is what is required.

Victorian and 'French' mantel clocks are characterised by the development of cases to a point where the dial as focal point of the whole may be obscured. Moreover, their movements were in very standardised form so that much of the antiquarian value of the clocks tends to be in their cases. Their aesthetic value is obviously a matter of individual taste, and the collector must beware of the wide range of reproductions and imitations which were made in the nineteenth century, but these clocks do not suffer from 'restoration' to the same extent as older wooden cases. Many a slate or 'marble' clock is a dingy black lump with a dirty bezel and a dial apologetically in the middle. If it is examined closely traces of old giltwork will often be found which, if restored, give it a totally different appearance which many owners relish, for it is often obvious that the amorphous dark mass is meant to be relieved by lightspots in symmetrical patterns.

Fortunately, there is not the same feeling among owners of these pieces that the dowdy and the dented are guarantees of authentic antiquity. Metal gilt-work can usually be brought up handsomely by washing in warm water and washing-up liquid, using a brush and even a brass-wire brush to penetrate the recesses. Rubbing with a paste of cream of tartar and water will sometimes brighten even further. Verdigris must be brushed or scraped out of an area, or it will spread. Such spots can be touched in with a gold wax or liquid leaf or, depending on the value of the clock, the fittings can be sent away for regilding, which is not practicable in the average home. A cast case which has been entirely gilt can first be washed and this may be sufficient, but it really is worthwhile having these cases properly gilt if the owner is prepared to meet the bill, for they will then be rejuvenated and good for very many years, and their sales value and market appeal will be greatly increased. At the same time, silks of the original colour (which will still be more or less apparent from beneath covered edges) should be fitted. It is remarkable how an extravagant metal confection can be transformed into an attractive, colourful – but not necessarily a brash 'new' – object by these simple means.

In all these clocks some attention should also be given to the bezels, which are usually polished and lacquered even in a gilt case. Particularly against a dark background of ebonised wood or slate, the cleaned bezel will restore the dial to its rightful prominence in the overall effect. White and black marble or slate cases can be brought to a good finish with a thin mixture of beeswax and turpentine, cracks and chips having been filled with painted plaster of Paris. Actual breakages can sometimes be mended by carefully drilling for pins, pressing together with a resin glue and covering the inside with a plaster of Paris support. Where the surface has minor chips it can first be levelled by rubbing with tripoli and water or turpentine. Metal polish, which is a very fine abrasive, is useful for cleaning white alabaster cases.

Carriage-clock cases are often very highly finished and the treatment may be as for gilt cases as above. Not all these cases, even the older ones, were originally gilt however and, especially on a plain unengraved case, a good lacquer finish may be more pleasing. Before cleaning, these cases must be completely dismantled. Access is almost always by undoing the pillar screws through the base, which may have a screwed cover, and generally the pillars are screwed also into the top. This is then in two or three parts held together by screws into the handle, which may be sprung into position and also secure the top glass panel. The older cases, however, are commonly in a single casting for pillars and top, with the handle pinned or screwed between brackets which are part of the casting, and the top panel also secured by pins or screws round a bezel. A gilt case must not be polished with metal polish, which will remove the gilding, but thoroughly washed and rubbed clean and dry. Lacquered metal should be soaked in a solution of household ammonia and soap or washing-up liquid, preferably hot, and then washed and treated with metal polish. The parts must be wholly immersed in the solution of ammonia, or high-water marks will be left which are very difficult to remove. Lacquer may be sprayed on to give an imitation-gilt dappled effect, but this spray lacquer is usually rather white in appearance, and transparent lacquer laid on with a brush is preferable. Lacquering must be done in a warm environment, and the pieces may themselves be warmed slightly to help the lacquer to go on evenly.

Flat and curved carriage-clock glasses can be supplied cut to size, and cracked and ill-fitting glasses should be replaced, because dust is a great enemy to these fine movements. The fit is especially important in striking clocks because a loose glass can cause a most irritating rattle when the clock strikes. Where the glass edge is thin in the pillar slots it may be wedged with a slither of cork or pegwood pressed in. Glasses should never be stuck into place with glue. A point to watch for is possible distortion of the case arising from a knock. If uncorrected, this can mean that even a glass of the correct size will not lodge in the slots of the crooked pillars but will tend to snap out or to crack when the case is screwed up. To avoid the same trouble, care should be taken that the pillars are replaced in their original positions with the steady-pins locating in their holes. Sometimes vandals have swapped the pillars and removed the pins in the effort to secure a better fit, and the best arrangement has then to be found and steady-pins knocked in.

Other points which often cause minor difficulty are the fit of the escapement index inside the door, the fit of the door itself, and the clearance of the repeater button with the front of the case, which can be a very close fit so that the repeater button stays down when pressed. This latter fault is usually caused by a bent angle-piece but, if necessary, metal can be removed from this or from the front of the case to give clearance. The back door is invariably hinged on pins and these are often bent, usually because the pillars have been inaccurately assembled at some time. The pins should be straightened, although in extreme cases it is sometimes necessary to take a little metal off the handle side of the door, for a sticking door is a source of irritation and can lead to the clock's being dropped in the effort to open it. The oldest carriage clocks tend to have solid brass doors, often engraved. Glass-panelled doors soon became general, however, and these are most often made with three sides in one piece and the front edge sliding in to complete the frame. This front edge is secured by screws into top and bottom and these can be troublesome, being subject to rust and breaking off in their holes. Then there is no choice but to drill them out and to tap for a larger screw, but it is essential to make sure that the countersunk heads are below the level of the metal, for they can cause the door to stick and also leave unsightly scratches on the case. It is in the interests of later repairers to ensure that, if a screw is filed off, a good slot is still left in its head.

Of the casing of contemporary cheap wall and alarm clocks there is little to be said. The movements are so priced that it is often more economic to replace them than to repair them, save for interest's sake, and the same goes for the cases, if the owner is concerned (which he rarely is). Snap-together cases wear after a period of use and it is rare that there is sufficient material for the raising of fresh locking nipples. A crack in a case can be repaired with fibre-glass or resin filler and thin metal cases can of course be soldered and bent if necessary. There is nothing that can be done to repair an anodised finish, but a good change in appearance can be wrought in painted metal and plastic cases by rubbing with a solvent such as some switch-cleaners. (The action should be tested on an invisible part first as sometimes ingredients destroy the surface.)

The glass domes of modern 400-day clocks are usually available from suppliers, though several may need to be tried. Some of these glasses will also fit the older movements. The larger glasses of the older types and of skeleton clocks can be a serious problem, and there is really no solution but to keep an eye open in junk shops and the like. Sometimes the cases of stuffed birds and similar relics can be used. It is possible to reduce the height of a glass by fixing a glasscutter in a vice at the required height and very carefully revolving the glass against the cutting wheel until the glaze is clearly broken, but it must be remembered that many of these glasses are extremely fragile.

Securing the Movement

The longcase movement is mounted on a stout seatboard running across the case, the 30-hour movement usually being unsecured if it is of the pillared type (though sometimes there are bolts through the bottom plate), and the plated movements being held by hooks round their lower pillars, passing through the board and fastened by nuts. Sometimes screws go through the board and up into the pillars. This is usual on old quality movements and may be one reason, apart from ornament, for the turned knop in the middle of the pillars. There are of course holes for the weight-lines and the many positions of these often suggest that movement and seatboard do not belong together.

The seatboard rests on the extended side boards, the cheeks, of the case. These are a common source of trouble – uneven so that the clock rocks, a favourite food of worm, and being chipped and split. It is not difficult to treat them for worm with a proprietary killer and then if necessary to screw additional boards inside and alongside them. Sometimes the seatboard is screwed to the cheeks, but whilst this makes for commendable rigidity it limits adjustment of the movement's position relative to the hood and makes removal difficult. If fastening is needed to prevent the movement from falling forward when the pendulum is removed, short pegs of steel, or bolts with their heads taken off, dropped into holes in the board and cheeks, seem preferable to screwing. The seatboard may go the full depth of the case, with a slot for the pendulum rod, or be little deeper than the movement, with the rod swinging behind it.

The movements of bracket clocks are customarily mounted on a small table at the front of the case, allowing clearance for the pendulum. According to the size of the dial, it may drop below the table, and these tables are often not fixed into their cases. In the older clocks, the fitting is most often through the pillars. But there was also an increasing tendency to use stout right-angled brass brackets, often engraved as part of the pattern on the backplate, screwed into the side of the case, even though this made the mounting visible, by virtue of the screwheads, from the outside. The conventional positions for these brackets were at the top left and bottom right of movements when seen from the back. This became the almost universal mounting from the middle of the eighteenth century until the demise of the heavy fusee bracket clock rather less than a century later. It was still used in clocks which might rather be called mantel clocks, being smaller and having a round bezel and glass rather than the heavy square door and brass or silvered dial of traditional form.

In the later nineteenth century what is perhaps the commonest modern mounting became general, these stout brackets being replaced by rather less stout lugs attached as a rule to screwed corner-pillars, rather than screwed into the backplates, and fastened into the front of the case alongside the dial rather than spanning the sides at the back. This method was of course more suited to the lighter movements being made and ensured that the dial was firmly in place against the bezel. There are usually four of these lugs and as a result the supporting table is often not used, the lugs taking the full weight of the movement. This mounting may be less than secure, especially when available space in the case has been used up with new holes as the old ones become too large for the woodscrews. The old angle-brackets, by contrast, received screws into themselves and were far less subject to wear. In the modern lugs it is often necessary to broach out the holes so that larger wood screws can be used. The situation is better when the lugs have nipples and holes, or angled projections alongside the plates, to keep their location constant, and sometimes they are mounted with an additional screw to the plate for the same purpose.

The movements of carriage clocks are almost invariably secured by two or four screws into the lower pillars, though many of the older ones, prior to say 1850, were secured by four small screws going into the thickness of the plates and angle brackets to the base were also used. The mounting of other French movements with their round plates and attached bezels is distinctive. There is usually also a rear bezel, whose door may be glazed or silk-covered, and the one-piece cases, which in themselves have no doors, are clamped between the two bezels by means of brass strips, usually riveted but sometimes screwed to the front bezel inside, and ending in right-angles with threaded holes. Long pointed screws go through the rear bezels and tighten up these strips so that the case is clamped between the two bezels. This is a simple and effective arrangement, but it has the disadvantage that the clock is fairly easily turned, especially by an enthusiastic winder, and set out of beat. On the other hand, over-tightening of the screws leads either to stripping of the brass threads or to breaking of the rivets, or to both. The straps can be bought, but in any case it is easy enough to make them, tap their ends (which are best bent over double for thickness of metal), and rivet them to the front bezel. Perhaps the most satisfactory repair short of replacement is to enlarge the hole and solder over it a good steel nut whose thread will not strip so easily. The hinges of these bezels, and still more so those of the older English mantel clocks with a large bezel secured by a latch in the side of the case, take a great deal of wear. They may require straightening and punching out of the old hinge pin, to be replaced by a new one, slightly tapered and filed off to shape. If, as often happens, one section of hinge breaks, there is little choice but to replace the whole, soldering on a new brass hinge as appropriate as possible.

The Dial

In general the dial of an antique clock gives more scope for improvement than that of a modern one, but it is very much a case of how far the owner wishes to go, bearing in mind the general effect. At the very least an old brass dial should be clean and legible. It will first have to be dismantled into spandrels (corner-pieces) and other applied ornaments, plate and chapter ring. The backplate is normally plain sheet brass with the centre, partly according to date, engraved or matted with dots from a multiple-headed punch, and the other parts are screwed in from the back. Again partly according to date, the winding-holes may be ringed (which was perhaps meant to lessen scratching from keys but became a fashion) or plain. Spandrels were in most cases cast in bulk and hand-finished and, providing they are consistent with other evidence, give some indication of date (for which the standard works, such as F. J. Britten's *Old Clocks and Watches* and H. Cescinsky's *English Domestic Clocks*, should be consulted). Generally speaking, the thinner and more complex the design and the poorer the finish (it is quite common to find swarf from casting not filed off), the later the spandrel is likely to be. The dating of chapter rings can also be carried out with some accuracy from their design. Chapter rings, which may now be polished and lacquered brass, were probably originally silvered, the figures and markings being deeply engraved and filled with black to form a superb contrast. Black figures on a plain brass background are never as clear.

The dial plate, like the plates of a movement, can be cleaned by soaking in a solution of ammonia and soft soap and finishing with metal polish. To secure absolute cleanness in the matted or engraved centre a small and soft brass scratch-brush is useful, and shaped pegwood will also help in clearing out the engraving. Paint stripper will usually loosen what ammonia fails to shift. The plate must be entirely free from smears and grease and can then be given two thin coats of clear lacquer applied with a soft brush. The flow of lacquer will be helped if the plate has first been warmed. Spandrels can be treated like the giltwork of cases, being washed with soap and brightened with a cream of tartar paste, but they should not be gilt unless they previously were so, because gilding can make them obtrude on the general appearance. They may also be cleaned with a brass scratch-brush and lacquered. Lacquer has to be applied sparingly to such pieces because a pool of lacquer has a dull, dark appearance.

Chapter rings should, in my own view, be resilvered if the silvering is worn and patchy, but the final finish is very much a matter of taste. A common and reasonable preference is that they should be nearly white in appearance rather than the shining silver which may result if a ring is sent out to be silvered professionally. They should not be given (for the first time) a deep circular grain with abrasive and there is no need to remove the accumulated chips and trenches which are characteristic of a really old dial. Resilvering can sometimes be

201

avoided if the trouble is in fact dirt rather than wear; rubbing with a stiff paste of cream of tartar and water will often resurrect the surface. Provided that care is taken with the figures, no great harm can be done by resilvering a ring yourself, and one of the traditional methods gives a very pleasant, not offensively new, finish. First, with emery paper of increasing fineness, remove all traces of the original silvering, rubbing dead straight in one direction (towards 12 o'clock) only. For the silvering you require silver chloride, which is a precipitate of silver nitrate in water; you may be able to obtain silver chloride from the chemist but, if not, soak silver nitrate in water with plenty of table salt and pour off the liquid, when purplish silver chloride will be left. Silver chloride degenerates rapidly if exposed to light, so keep it covered. Now wet the cleaned ring thoroughly with salted water and rub over it thoroughly with a moistened rag charged with the silver chloride, when the ring will turn a dirty brown. Go over the ring with a paste of cream of tartar and this brown will turn to whitish silver, patchy at this stage. The work should now be washed in very hot water and dried, preferably by waving in the air or in front of a hair-drier – the object is to avoid finger-marks. If there remain some brown streaks, rub them again with cream of tartar paste and repeat the process until all is even in colour. The ring must then be warmed sufficiently to cause the figures to shine, but not, of course, to such an extent that they boil and dry dull and cracked. This warming will help in the application of the thin coat of lacquer which must follow if the silvering is to last. Damaged figures can be touched in with a pen and lamp-black oil-paint thinned with linseed oil, once the lacquer has dried; and it should be noted that this paint takes several weeks to harden. Strictly, the figures are filled with black wax before silvering, but the wax is not easy to obtain, though sealing wax is excellent if available. Black boot-polish mixed with clear lacquer is also effective but hard to apply – it will require trimming when set. It is easiest to reduce all these pigments with suitable solvents so that they can be applied with letter-writing pens; these will stay in the right place even if the colour is harder to build up than with a brush.

The entirely silvered dial, usually without separate spandrels and chapter ring, came into fashion in the middle of the eighteenth century.

It is no easy matter to resilver consistently so large an area of metal, and the figures and markings are usually engraved much more shallowly than in the older chapter rings, which makes them difficult to touch up. If the dial is in such bad shape that it must be resilvered, this work is best given to a specialist. Do not, however, admit defeat until an attempt has been made thoroughly to clean the dial with a soft cloth and the usual cream of tartar paste, for this will often revive the colour considerably once the lacquer has been soaked off. The markings can be touched up later with a pen so far as possible. Sometimes, of course, they are gilt and liquid leaf will need to be used.

The restoration of painted and enamel dials is very difficult. The markings are not engraved but superimposed on the white surface and they have therefore to be touched up virtually freehand which, even if the old line half shows, is chancy on a glossy surface. Moreover, these markings are not always stable and it is unwise to attempt to clean the dials with ammonia or a mild abrasive like metal polish, at least until an obscure and minute corner has been tried. An old standby where simple cleaning is involved is a lump of doughy bread, sometimes with milk. Special dial rubbers for the purpose can also be bought. There is very little that can be done with a badly chipped or cracked enamel dial save to replace it. A small chip or crack can be cleaned out with fine emery and filled with a thin china cement such as 'Porcelainit'. The surface of a large painted dial, which is usually on iron, is often crazed with minute hairline cracks which, if they do not detract from the appearance, it may be a pity to blot out. There are specialist firms who will repaint and redivide these dials, but one has to beware of the somewhat cheap and bright appearance which, whilst it might be historically authentic, may not be desirable. Much can be done to the paintwork, where the outline is fairly clear, with poster paints, which have the advantage that errors can be wiped off, though the dial will afterwards have to be lacquered to stabilise the markings.

Enamel carriage-clock dials are sometimes interchangeable if the 'feet', which are pinned through a brass subplate in the better clocks or through the front plate in the later ones with motion work between the plates, can be moved. Here it is necessary to scrape carefully through the back enamel to the underlying copper so that a moved foot, or a new one made from heavy copper wire, can be soldered on. The dial must be well-supported, for these dials craze very easily. When fitting a new foot made from wire the job is simplified if the wire is not cut till after the job is done, but is used in a free-standing curve with the end falling onto the plate at the right place. Some cracks can be cleaned and filled or, if the owner agrees, covered by a brass or copper mask with holes for dial rings. Unless a suitable mask is available (for some clocks were fitted with them from the outset), the home-made mask is best gilt, since it is not possible to make on it the engine-turned patterns which are found on the originals. The alternative to this sort of work is to buy a photographic reproduction dial to stick over the old one, or to the brass subplate alone if there is one. This is in some ways an admission of defeat, but a good-quality reproduction with its glazed surface is in some cases preferable to leaving a severely blemished dial. Care has to be taken with the fitting of the hands if one of these dials is fixed over the original, because the extra thickness may mean that longer collets have to be fitted to hour and alarm hands for the hands to be secure and not to rub the dial.

Of the mass of modern dials it can only be said that little can be done but to replace them. It is usually simple enough to obtain a dial of appropriate size and to drill or punch it with winding holes as necessary, but it is not always possible to obtain a dial in the right size with comparable figuring. The characteristic 'silvered' dial of the recent chiming clock cannot be repaired if scratched or worn, and neither the metal nor the appearance will permit traditional silvering processes to be used. Moreover, the figures are of course not engraved. These dials can only be replaced or cleaned and left otherwise as they are. As to the mounting, where fold-over lugs are concerned, one has to improvise, since the metal is often such that new lugs cannot be soldered on, though resin glue may be used. In extreme cases an aluminium subplate with lugs can be fixed to the dial or small screws can be used behind the bezel. In many instances the mounting of the movement in the case will, in fact, secure the dial reasonably well without the need to replace lugs.

Mention should be made here of the so-called 'digital' clock. There is nothing very novel about it in general, digital indication being an all-but universal feature of, for instance, meters and the dashboards of cars and having been in use in the calendars of clocks and specialised (particularly night) clocks for several centuries. But the use of numbers in an aperture, rather than an index pointing out one of a series of numbers visible, is a contemporary fashion for which, no doubt, many explanations could be offered. Suffice it to say that the system gives no visual indication of the context of the moment, but indicates the moment only.

Digital indicators nowadays are of two types, the continuous cylinder and the flip-over number-plate. In addition there is, of course, and always has been, the segmented circle. Both main systems require the placing of wheels, or at least motion wheels, in a different plane from that normally used in a clock and both are particularly suited, therefore, to the compact electric motor which does not have to be wound and whose armature and dependent gears can revolve parallel to the dial. So far as the cylinder is concerned it is simply a matter of connecting the appropriate toothed wheel to a cylinder with numbers disposed round its edge, or making it entirely of one plastic moulding. In the digital revolution-counter a pin or peg pushes round the 'tens' once for each revolution of the 'units', just as the pinned day-of-the-month wheel of a calendar clock pushes round the name-of-the-month starwheel at every revolution. Thus, for this period of engagement the wheels and digits revolve at the same speed, but there is no continuous connection between them. A similar arrangement is adopted for the double figures in the 'cylinder' type of digital-clock dial, the minutes reverting to 'nil' as the hours are advanced one digit, and the 'unit' minutes repeating a cycle as the 'tens' are advanced up to 'five' (fifty).

The 'flip-over' system, though it appears to move only on the minute, is nowadays usually driven by a continuously revolving arbor with full gear connections. Each number is held in a pile on the arbor, the edges being tapered by being constrained into a circle where they are hooked onto the arbor, and released from a restraining hook as the tangent is passed (Fig 59). In practice the numbers are cut in half and displayed as an open book, the bottom half of one number having on its reverse the top half of the number before. These numbers may not be in a separate series for 'tens' and 'units' but, being so compact, be

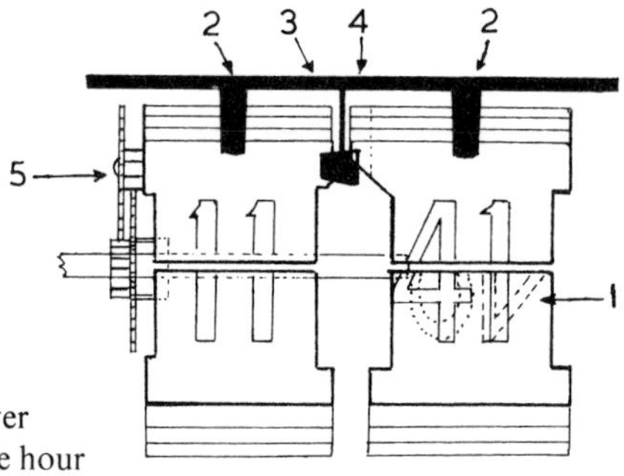

Fig 59

Flick-over card digital display
1. Reverse of lower half card bears top half of previous digit (shown dotted)
2. Card retaining clips
3. Special retaining clip moved over in second half of hour to ensure hour card does not fall until minute 60
4. Sloped edges of later minute cards operate special hour retaining clip
5. Minute wheel and pinion, driving hour wheel on whose pipe the hour cards revolve

in a complete series from 00 to 59, so far as the minutes are concerned. The stack of numbers is arranged on a ratchet so that it can be manually advanced but the arbor cannot be turned backwards. The hour stack is connected to the minute arbor by conventional motion wheels so that one revolves once for twelve revolutions of the other. Sometimes the hour wheel is on a 24-hour cycle. The hour stack has a sideways catch which prevents the next hour falling, in addition to the usual tangential catch. It is so set up that this catch is displaced at about 45 minutes past the hour, by extensions to the minute stack plates, so that when the tangential catch is passed by the hour plates, the next hour cannot appear until the last of these extra-wide minute plates, representing 59, has fallen. Thus the hour plate is, as it were, held at 'warning' for the last minutes of the hour, having already been released by its own catch but being held up by the minute stack. In some clocks the hour stack has two sets of hours on it for each hour, the plate changing, but the number staying the same, half-way through the hour. This is merely a means of filling up the hour stack. As a matter of interest, flip-over clocks were, before their modern vogue, mass-produced in Germany in the early 1900s as digital carriage clocks with visible mechanism, brass cases and glass panels. Their principles were similar in those just described.

The Hands

The hands of old clocks are fairly reliable indications of date if they can be assumed to be original. (The reader is referred to the books mentioned on page 201, for details.) They were normally hand-finished, or filed up completely by hand, and made of blued steel. The method of mounting depends on the mechanism of the clock concerned. Where the motion work involves a cannon pinion or wheel sprung from behind by a washer on the front plate, the minute hand has a square hole to fit over the squared end of the cannon wheel's pipe. This hand must be correctly positioned over the lifting pins of the striking, so it is normal for a round brass collet with a square hole to be riveted into the boss of the minute hand. If, when the hand is as well-positioned as possible, the striking still occurs just after or just before the hour, this collet may be moved slightly by putting it on a square pin in the vice and gently moving the hand from the boss (*not* by the tip, which is liable to break the hand). When the correct position has been found, the collet should be riveted tightly into the boss with punch and hammer, especially if time-setting is by moving the hand rather than by set-hands arbor. (If the hand has no brass collet and such an adjustment is still required, it must be made by bending or, in severe cases, moving the lifting pin, which should not be necessary if the hand has always belonged to the clock. If the lifting pins are on the minute wheel rather than the cannon pinion itself, move the hand by adjusting the relative position of minute wheel and cannon pinion rather than by altering the hand.) The hand in this arrangement is held to the cannon pinion by a pin through the end of the arbor and with a domed collet over the base of the hand. A similar set-up is used for French clocks with the split frictional cannon pinion, where if the hand is not pinned on with sufficient collet the cannon pinion's lifting pins may fail to engage the lifting piece and the clock will not strike. The pin must be such as to fit the hole in the arbor, but the choice of the correct collet is, as has already been said, essential if the right tension is to be kept on the hand for running and for time-setting; time spent on this fitting is never wasted.

On clocks where the time-setting consists of a sprung central arbor and hollow centre wheel, often with motion work between the plates, the minute hand has to be fixed direct to the centre arbor, on which may also be attached the lifting cam for setting off the strike. Sometimes a round brass collet is riveted to the hand and the whole is driven onto the arbor. Often the end of the arbor is squared and also threaded and the hand is secured by a screw rather than a pin, or the hand is punched and drilled through so as to leave slightly thickened metal and the plain hand is pushed on. The brass collet's hole can be tightened with a punch and the end of the arbor can be roughened to increase grip, but there is little that can be done to improve the grip of a straight press-on hand once it has given trouble, save to cramp the metal in a vice or with a hammer. One is then inclined to go too far (for the fit is critical) and to have to resort to broaching the tightened hole out, thereby removing valuable metal, and the last state may be worse than the first. When such a hand has been on and off a few times the hole ceases to be true and the only cure is replacement, which, with a modern hand, is not usually too difficult. The same considerations apply to alarm hands (whose setting up was mentioned at the end of Chapter 5). They must be accurately placed and driven on firmly, most easily with a hollow punch.

The hour hand of an old single-handed clock is pinned, often without a collet, to a squared arbor pivoted between dial and front plate (Fig 33), there being no central arbor in the usual sense in these clocks. On this arbor is the hour wheel, engaging with a pinion on the driving-pulley arbor, and pinned to the 12-toothed starwheel which lets off the striking at each hour. The wheel and starwheel are held on the short arbor by a round friction-washer riding in a notch in the arbor. Where there are two hands and a centre wheel the hour hand is mounted on the pipe of the hour wheel. In old English longcase and bracket clocks the hand may be square-holed and the pipe also be squared, in which case the pipe is often clipped friction-tight to the wheel so that its position can be adjusted. Whether or not the pipe is squared (or has a locating flat on it) the hand is usually screwed into the edge. The striking snail may be solid with the hour wheel or screwed to it, in which case care has to be taken to mount the hour wheel so that the hand points to the hour with whose section on the snail the rack-tail will engage. Where the countwheel system is used, or where the snail is frictionally mounted or on a starwheel which can be turned independently of the hour hand, this problem does not of course arise. On later and continental clocks, including carriage clocks (which, however, often use the snail on a starwheel), this adjustment can nearly always be made after assembly, for the hour hand is not screwed on but mounted on a split collet, merely pressed over the hour pipe and held against it by friction. Within limits the fit can be tightened by compressing the collet, but if too large a hand is used only the tip of the collet will grip the pipe and the hand will be insecure.

Centre-seconds hands are usually pressed onto the end of their fine arbors. Particular care has to be taken to see that these arbors are not bent. The hands are under no strain and their fit can be tightened if necessary by nipping the collets, subject to the same limitation as with hour hands above. A special point to watch with all concentric hand arrangements, but particularly the fine ones used in the last 150 years or so, is the need for clearance between the pipes, between the hour pipe and the dial, and between the reverse of one collet and the front of another. Rubbing at any of these places can be adjusted with a fine file, but it is often the cause of an obscurely stopping clock and nowhere more so than in a clock with centre-seconds whose seconds arbor is imperceptibly bent. Bending can be caused by rough treatment of the hand, but is usually due to laying the movement on its face to remove the hammer, suspension cock, and so forth. It is wise to deal with these movements mounted on a box or jar or else to tackle the backplate with the movement standing up.

Hands (with pronounced tips used for setting) for single-handed clocks are often not available, but it is possible to adapt an hour hand of later type by filing up a tail and fixing it with hard solder. The hands of longcase and bracket clocks can in fact be cut from sheet metal like fretwork (that is, drilling the corners before sawing) and filed to a finish. It is a long job, but it makes for a more convincing article than modern replacements. Brass and steel hands that are broken can usually be silver-soldered together, though it is best for both parts to be held in position with an adjustable double clamp (which can be bought) for this work. Soft solder will not be strong enough save for ornamental work of no structural importance.

Replacement hands are available for most types of clock, but they tend to look like replacements, being stamped out sharp and symmetrical, and are better doctored before use. The other difficulty is that it is rarely possible in practice to obtain, at the right time, both the correct fit and the correct style, especially for old hands where the period style is important. For example, the longcase hands widely available new are mainly of the later serpentine shapes which look grotesque on a simple old dial, and French- and carriage-clock hands may only be available with solid spade points when really it is the hollow-moon type (of a distinct period) that is required. If the right style is available, it is best to buy it and to adjust the collet to fit, or attach a new collet. Depending on the value of the clock, it is better to replace a pair of hands than a single one, unless the second one really is a partner to the survivor. Often an old collet can be punched off a broken hand and riveted onto the new one. Sometimes it is possible to buy a long, plain modern hand and to shorten it – an anonymous plain hand may be preferable as a temporary measure to an obviously modern reproduction in the wrong style. For nearly contemporary clocks one is, of course, less cramped. Provided that the hands are of the right length (preferably that the minute hand does not extend beyond the outer ring and the hour hand's tip does not pass beyond the feet of the numerals) owners are often not fussy as to style, though they will naturally require luminous hands with a dial. Luminous paint can be bought for touching up damaged hands and dials.

Old steel hands are normally blued, and indeed blued steel hands will often look better even where plain brass ones, always rather indistinct, were originally fitted, if the proper replacement is not available. The quick method for the coarser hands is to paint with blue-steel enamel; and this can be effective provided that the coat is kept thin and even, which means keeping the hand strictly horizontal until dried. It is always better, however, to blue with heat. Get the metal perfectly bright and clean, removing especially all rust with emery and oil, and cleaning off with benzine, and then do not handle it with the fingers. Place the clock hand on a blueing pan – a small brass container filled with brass filings to distribute heat and with graduated holes for hand bosses – and heat it. Alternatively, use a suitable piece of thick brass with a hole, such as an old clock-plate, for the hand must lie flat in contact with solid metal or it will not be evenly heated. In applying heat, remember that the thin point and shaft tend to heat more quickly than the thicker parts. When the desired shade of blue has nearly been reached, remove the hand with tweezers or pliers, plunge it into fairly heavy oil and, when cool, remove it and wipe with petrol or benzine. Two or three efforts may be needed to produce an equally blue pair of hands, but blueing with heat achieves a far gentler and more traditional colour than treating with enamel.

Finally, whether or not a new hand is bought, do take care that the hands are a pair. Frequently one receives carriage clocks, for example, where the minute hand of a small French clock has been adapted to match an existing hour hand. This is a legitimate combination, but the hand from the larger clock cannot simply be cut down and fitted; it must usually be made very much finer as well or it will be horribly out of proportion. The public often regard fitting and adapting hands as additional jobs costing neither time nor money. 'Oh, and it needs a new hand,' is a frequent afterthought. In fact, if hands are seen-to properly they may take as long as all the other repairs put together, though whether one charges accordingly is a personal matter.

Hands are the parts of a clock which are seen first and most. If they do not 'belong', or if they come loose, no amount of fine workmanship between the plates will silence the complaint. This includes the small 'index' hands of the strike/silent and regulator type. They are not difficult to replace and indeed are usually so strong as seldom to need replacement, though one may be missing. But they must be firm on their squared arbors, pinned home decently with a new pin protruding equally on either side, and so placed that the cams against which they usually operate do not reach their stopping-point when the indicator is only half-way round the dial. As for pins — new pins should always be used for fastening hands, whatever economies may be made elsewhere. They make a world of difference to the appearance. These external details count for a great deal and they are an almost daily irritant if incorrect. They are also an important factor by which, assuming that the clock gives no actual trouble, the repairer is judged.

8 THE REPAIRER AT HOME

Clocks old and new are of enormous variety, and it has not been possible to cover a full range in detail. There are, however, certain common principles and systems – the driving and the driven wheel, the oscillator and the escapement, the rack and the countwheel or the slotted wheel of alarms – and there are also common points of adjustment and repair associated with those principles. It is hoped that concentration on these will assist the repairer in many circumstances, so that he can appreciate the details of movements before him and will both recognise departures from common practice and be able to devise sound means of dealing with them.

Now, more than ever, departures from the norm are not made without reason. We live in an age of mass-produced movements in international markets. It is not economic to introduce novelties almost for their own sake. Novelties must appeal to the sophisticated market by being an improvement in time-keeping or convenience; and to the public at large by being cheaply replaceable without costly servicing even if the individual item has no great durability, or by being attractive, even sensational, in appearance, coupled again with convenience. The repairer should, before starting work, have clear in his mind the apparent reason for what seems unusual. He comes to detect the novelty which is radical, requiring different methods, from the novelty which is superficial only, and he works accordingly.

It may fairly be wondered whether the small repairer of today has any future. The public's requirement of a clock has changed over the years. It expects (perhaps partly because of the example of the often maligned synchronous movement) accurate time-keeping over a long period with the minimum of attention, and it can check this time-keeping with ease by radio or telephone. It expects (sometimes unreasonably, sometimes not) something not far short of this standard in its antiques also. Nor do people expect to spend much on buying a clock, save perhaps as a wedding present or in setting up a new home. The clock is looked to less as the master-timekeeper, to be deliberately consulted, than as the reliable visual reminder available, very likely, in every room of the house. We take the clock on trust and it has become a pattern of our behaviour to give it a glance and yet, if someone subsequently asks us the time, we have not registered what the clock said. It is a minor jolt to the system when the face which we treat so discourteously becomes distorted and manifestly wrong. Increasingly, when the clock does not adequately perform in this curious ritual, it is simply replaced by a more amenable one which will do what is asked of it, such is now the relative cheapness of clocks keeping reasonable time. In this the retailer, who wishes to sell a clock and knows the cost in real terms of repairing any clock today, naturally plays his part. The local shop, depending as much on repairs done on the premises as on pieces sent down from head office, is increasingly a rarity, and its attention is largely concentrated on movements no longer in production, those old enough to have sentimental value to their owners and, maybe, the genuine antique.

In this situation there is in fact a great deal of scope for the amateur, provided that he does not tackle and bungle things far beyond his level of experience. Again and again he can redeem pieces which have been declared beyond repair, whether they are ancient or modern. The scope is attested by the large number of requests for help which one will receive even without any effort or advertisement on one's own behalf. The simple fact is that the amateur home-repairer is himself prepared to pay, with his interest and his love of working with clocks, for a proportion of the expensive time and overheads which put many a professional repairer out of business or lead him to favour the more predictable margins to be obtained from selling. This is, for the home man, an account which generally balances. At the same time, he has to be wary that he is not exploited, for this can be very dispiriting. If one is doing repairs for other people, sooner or later the question of reward will have to be considered. Charging for repairs is one way to moderate excessive demands and to appease one's conscience as to time spent. More important, many people value more highly work for which they have paid, however nominally, and this in turn tends to breed a self-respect which will improve the work of the repairer.

Everyone has his own solution, and there is no particular merit in consistency. One may accept an ounce of tobacco or a box of chocolates or a token sum 'for the children', or in other cases refuse to accept anything. If, of course, one wishes to claim tax relief on expenses and to enter into competition with professionals by running a business, that is another matter. I am thinking rather of the clock-lover who from time to time provides a service which the professionals of the area are known not to provide. A popular compromise is to charge for parts. This has its snags. Inevitably one buys in a supply of parts, such as suspension springs and even some fairly costly 'repeat' items like platform escapements, for the future or when a cheap line is available. In due course one forgets their unit-cost or loses track of the current price. Moreover, the cost of a new screw can be very insignificant compared with the time spent in extracting the old one. To charge for a single screw may appear miserly, yet the original package of several gross may have been paid for with a not inconsiderable cheque. These costs mount up, and it is a question as to how much one proposes to spend in the name of one's hobby (and personal pleasure) for the benefit of the public more or less at large. If labour is charged and a profit made on parts, one starts to run a business, and this has complications (despite the benefits) which many amateurs would rather avoid.

These are matters which only the individual can decide in the light of his own circumstances. If it is decided to charge, however, by whatever method, it is a sensible rule always to give an estimate, making sure that the estimate will cover all conceivable works and parts, though not accidents. One must expect to pay for these oneself, the risk being assessed when one accepts the job. You are not (unless you choose to be, and it is something to be discussed on each occasion) in competition with the professional, and you have no need to pride yourself on a cut-price job. You prove yourself by sound work and the unquestioned right of the owner to return to you if necessary. He, if he can have the job done professionally, insured and guaranteed, should know what he is doing if he comes to you instead. It is far preferable in such circumstances to give a high estimate, with the chance of its being turned down, than to give a low estimate to which you must stick and then sustain a loss and possibly do poor and hurried work. There is no surer way (short of utter incompetence) to ruin a job than to be committed to doing it to a schedule and to a price both of which turn out to be unrealistic.

There is no necessary conflict between daily work, in the office for example, and spare-time work on clocks for others. If the latter becomes something of a business, however, there is certainly a potential conflict and it is necessary to watch for the situation where pleasure becomes work and the day becomes one long extended grind. The amateur is under personal obligations only, working out what is within him, helping those who seek particularly his help, able to turn down a request because his bare livelihood does not depend on it. For such a person there will always be opportunities ranging from the adjustment of the cheapest alarm clock to the occasional investigation and restoration of a noble antique, and the present tendency of the industry towards the disposable and replaceable can only lead to an increase in such opportunities. I hope that those so favoured may be helped by this book.

FURTHER READING

Horological literature has grown greatly of recent years and a comprehensive bibliography would not be practicable here even if it were appropriate. The following is a selection of the books which I have found most useful. The two sections roughly indicate a division of interests between general, historical and antiquarian books written partly with collectors in mind, and practical and mechanical books written for repairers and restorers. The division is not, of course, complete, and much mechanical information will be found in books listed in Section A, as well as some historical information in Section B. Where a book serves both interests substantially it is included in both sections.

A   Descriptive and historical books

Allix, C., *Carriage Clocks, Their History and Development*, Antique Collectors' Club, Woodbridge, 1974

Baillie, G. H., *Watchmakers and Clockmakers of the World*, N.A.G., 1929, ed. 1974 (*see also* B. Loomes)

Bird, A., *English House Clocks 1600–1850*, David and Charles, Newton Abbot, 1973

Britten, F. J., *Old Clocks and Watches and their Makers*, see C. Clutton

Bruton, E., *The Longcase Clock*, Arco, 1964

Cescinsky, H. and Webster, M. R., *English Domestic Clocks*, Routledge, 1913, repr. Hamlyn, Spring Books, 1969

Clutton, C., Baillie, G. H., Ilbert, C. A., *Britten's Old Clocks and Watches and their Makers*, Eyre and Spottiswoode, 8th ed., 1973

Cumhaill, P. W., *Investing in Clocks and Watches*, Barrie and Rockliff, 1967 de Carle, D., *Clocks and their Value*, N.A.G., 1968

Edey, G. W., *French Clocks*, Walker, N.Y., 1967

Edwardes, E. L., *The Grandfather Clock*, Sherratt, Altrincham, 3rd ed., 1971

Goodrich, W. L., *The Modern Clock*, North American Watch Tool Supply Co., 1905, repr. 1950

Jendritzki, H. and Matthey, P., *Repairing Antique Pendulum Clocks*, Edition Scriptar, Lausanne, 1973

Loomes, B., *The White Dial Clock*, David and Charles, Newton Abbot, 1974

———, *Watch and Clockmakers of the World*, N.A.G., 1976 (supplement to G. H. Baillie, *see above*)

Robinson, T. R., *Modern Clocks*, N.A.G., 2nd ed., 1942

Royer-Collard, F. B., *Skeleton Clocks*, N.A.G., 1969

Smith, E., *Repairing Antique Clocks*, David and Charles, Newton Abbot, 1973, repr. 1975

Terwilliger, C., *The Horolovar 400-Day Clock Repair Guide*, Horolovar, N.Y., 1965

Tyler, E. J., *European Clocks*, Ward Lock, 1968

Ulyett, K., *In Quest of Clocks*, Barrie and Rockliff, 1950, repr. Hamlyn, Spring Books, 1968

B   Practical books

de Carle, D., *Clock and Watch Repairing*, Pitman, 1959

———, *Practical Clock Repairing*, N.A.G., 1952

Britten, F. J., *Watch and Clockmakers' Handbook*, ed. from *c.* 1870, 11th ed. repr. 1976

Darnall, J. V., *Restoration of Wooden Movements and Cases*, Tampa, Florida, 1970

Fried, H. B., *Bench Practices for Watch and Clockmakers*, Columbia Communications Inc., 1954, repr. 1974

Gazeley, W. J., *Clock and Watch Escapements*, Heywood, 1956, repr. 1975

———, *Watch and Clockmaking and Repairing*, Heywood, 1965, repr. 1975

FURTHER READING

Goodrich, W. L., *The Modern Clock*, North American Watch Tool Supply Co., 1905, repr. 1950

Harris, H. G., *Watch and Clock Repairs*, Arco, 1961

Jendritzki, H. and Matthey, P., *Repairing Antique Pendulum Clocks*, Edition Scriptar, Lausanne, 1973

Randell, W. L., *Clock Repairing and Adjusting*, 1923, repr. Percival Marshall, n.d.

Robinson, T. R., *Modern Clocks*, N.A.G., 2nd ed., 1942

Saunier, C., *Treatise on Modern Horology*, eds. from 1888

Smith, E., *Repairing Antique Clocks*, David and Charles, Newton Abbot, 1973, repr. 1975

ACKNOWLEDGEMENTS

I should like here to express my sense of indebtedness to previous writers on these matters. They are too numerous to name and I may well be unaware of the fullest extent of their influence, as they have helped me over the years. I am conscious also of a similar great and general debt to very many clockshops and antique shops who have, often unwittingly, enlarged my experience of clocks, and also to friends and customers with whose pieces I have had to deal. Here I would specifically mention the Reverend David Clift, who kindly let me photograph his Comtoise. My biggest and most immediate thanks are to Dr and Mrs A. J. Allnutt for their photographical expertise, and to my wife for her long-standing toleration of things horological at all hours of the day and night.

INDEX

222